Servicing Cassette & Cartridge Tape Players

No. 716
$9.95

Servicing Cassette & Cartridge Tape Players

by Homer Davidson

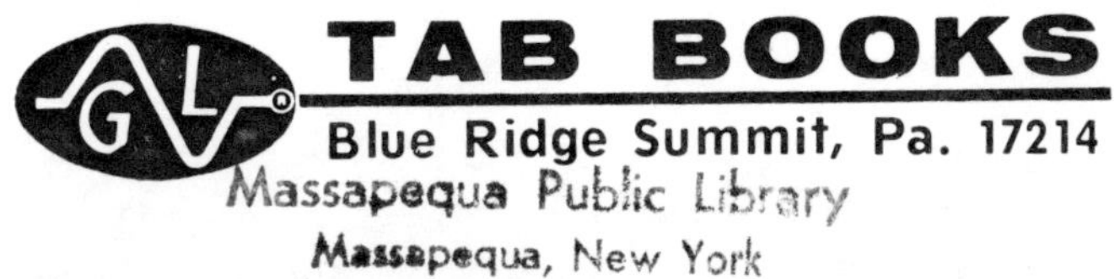

TAB BOOKS
Blue Ridge Summit, Pa. 17214

FIRST EDITION

FIRST PRINTING—JANUARY 1975

Printed in the United States
of America

Hardbound Edition: International Standard Book No. 0-8306-4716-3

Paperbound Edition: International Standard Book No. 0-8306-3716-8

Library of Congress Card Number: 74-25564

Preface

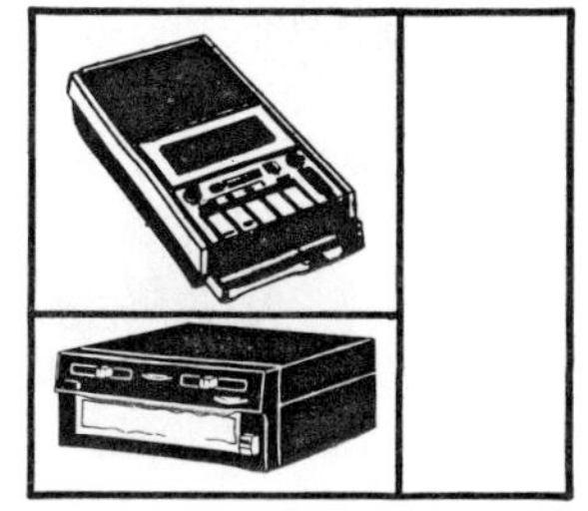

There are several thousands of cassette and 8-track stereo tape players in use today; this means big money volume for the servicing technician. Many of the large U.S. TV manufacturing companies have added the cassette and 8-track tape machines to their production lines. One recent survey indicated that the average family has no less than four tape players of one kind or another—which makes the tape player more ubiquitous then the TV. Who services these units when they need repair? Many TV repair shops have found tape player service a lucrative sideline. Several well known TV repair depots have turned to tape player repair as a secondary specialty service. You too can service tape players—either as a specialty or a sideline. Once you master the basics, it's easy. But more important: it's profitable.

The purpose of this book is to show you how to get started servicing tape players. Actual testing, adjustments, and troubleshooting methods are included in each chapter. This is not a book of theory; rather, it's a practical source of information applicable to real-life cartridge and cassette players of every description.

The first six chapters deal with types of players, circuits, testing and adjusting, lubrication, cleaning, and speed problems. Actual troubleshooting of the individual tape machines are given in the next five chapters. Actually, you can begin to service these tape players by just reading these five chapters. Troubleshooting charts, included at the end of each chapter, may be used to help you rapidly pinpoint possible troubles. Chapter 12 includes actual representative case histories. Although some models may not be found in your area, principles will prove universally applicable, and the problems are typical.

Many companies have furnished servicing data and circuits for this book. The extra help and cooperation of manufacturers' national service managers have been excellent. Without this kind of help, this book would have taken years to prepare.

I gratefully acknowledge the cooperation of Admiral Corp., Allied/Radio Shack (Tandy); Automatic Radio, Channel Master Corp., Craig Corp., Gambles, Hitachi Sales Corp., Lafayette Radio; Lear-Jet Corp., Lloyds Electronics, Masterwork Columbia Records, Midland (Microtron), Montgomery Ward, Nivico-JVC, Matsushita Electric Corp. (Panasonic), RCA Consumer Electronics, Sharp Electronics Corp. and Sony (Superscope) Corp. Technical data and tape player circuits of the various companies are found throughout the book. These companies are eager to help you service their products.

Homer L. Davidson

Contents

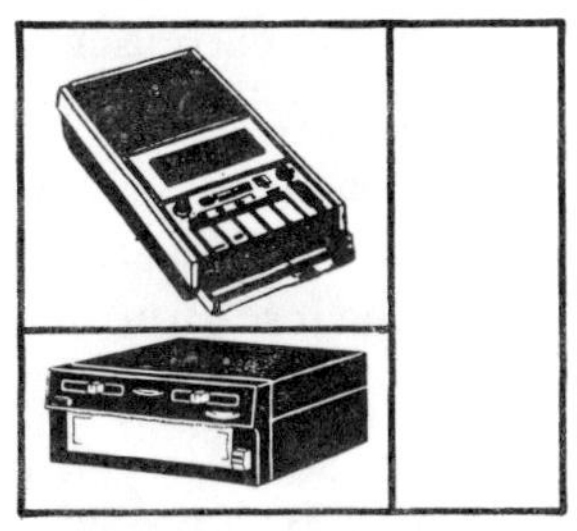

Getting to Know Cassette and Cartridge Players

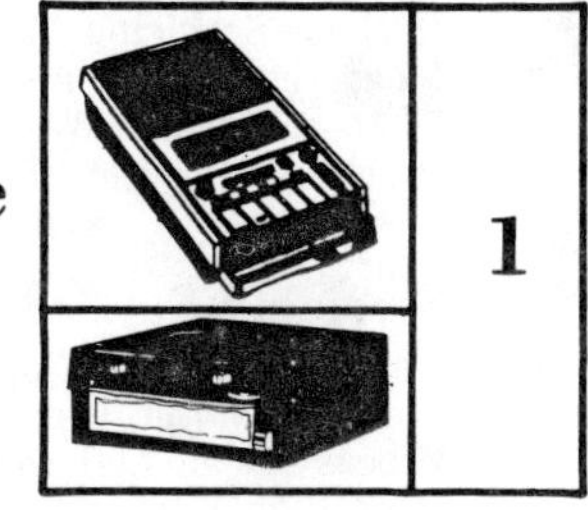

1

Do you know that about every major or small consumer-electronics manufacturer produces a cassette or cartridge tape player of some type? Don't believe it? Just take a look down the street at all of the electronic, TV, appliance, drug and discount stores selling this merchandise today. The majority of these units may be manufactured in Japan for or by even the largest electronic manufacturer. There are millions of these units in service today and each year the entire consumer-electronic market grows at the staggering rate of 25%!

Where does this leave you? Service—yes, any electronic unit must have service—and these small tape player repair tickets can add up fast. If you are not servicing tape players, you've been missing a very lucrative source of income. You can keep several men (yourself, too) busy servicing these tape machines once you get serious about it. When the slow season for color TV service rolls around, what could be more logical than to keep busy servicing cassette and cartridge tape players? And there's a bonus: Most of these repairs are of such a minor nature as to be easily sandwiched between larger jobs.

Circuit simplicity in tape players, coupled with a high degree of interproduct uniformity, means easy repair. You may want to choose several large manufacturers and do the warranty service on their tape player line. You can always get replacement parts this way.

In case you are now servicing solid-state auto radios, the auto tape player is right down your service bench. Yes, now stereo and quadraphonic tape players are even found in the stereo radio units. Some of the new color TV receivers have tape players in them, too. A little applied knowledge and a

Fig. 1-1. The cassettes; 1. Cassette test tape; 2. Torqette torque cassette; 3. Recording cassette; 4. Prerecorded cassette.

dash of common sense are ingredients that add up to easy profits in cassette and cartridge tape player repairs. Let's get started by discussing cassettes and tape cartridges.

The Compact Cassette

The tape cassette is a complete tape-handling system in miniature! Within the cassette are the supply and takeup hubs, which move the tape in either direction. The cassette tape is not an endless tape loop like the 8-track cartridge. The tape package contains the necessary guide rollers, pressure pads, and reels. Cassettes are available to play for 30, 60, 90, 120 and 180 minutes. Typical "quality" cassettes are pictured in Fig. 1-1.

The cassette operates at 1⅞ inches per second. The tracks are unique in that they are adjacently positioned. Cassettes for home use have a frequency response extending from 40 to nearly 20,000 Hz. The equipment used to play the cassettes has been improving steadily in recent years, too. A high-fidelity cassette system today will perform as well as a disc playback system of comparable quality. Wow and flutter with the cassette is less than 2% in good tape players. Noise—the traditional "big" problem with a narrow tape track—has been improved with better tapes, higher bias levels, and the proliferation of Dolby and other well designed noise reducing systems.

The cassette must be protected from environmental extremes of heat and humidity. A tape cassette should never be exposed to direct sunlight for any appreciable period. Similarly, tapes of less than exceptional quality must be kept in a dry, dust-free, and fairly cool environment. Otherwise, the listener will be exposed to such troublesome and annoying problems as playback squealing, tape sticking, and erratic speed. Always recommend high-quality cassettes to preclude callback complaints along these lines.

One feature unique to cassettes warrants mention for the benefit of service specialists who might be unfamiliar with this reproduction medium. The cassette is intended to be "goof-proof"—and to this end, cassettes have been designed so that the inadvertent recording over prerecorded tapes can't happen. Each blank cassette has a couple of knockout plugs on the spine; when these have been removed, the associated recording equipment cannot impress any new signals on the tape.

As you begin to service cassette recorders and players, you'll probably find the most popular complaint to be *"it won't record."* Nine times out of ten the problem can be traced to these knockouts—users just seem to forget they exist. To restore a tape to recordable condition when this complaint comes up, just place a small piece of transparent tape over the knocked out panels on the rear of the cassette.

Another problem that rates high in terms of repeat performance is fouled tape. Cassettes have improved so drastically in recent years that there is very little excuse for fouled tape unless the equipment or the cassette itself is of very poor quality. (Even "good" tapes can foul when a heavy-handed operator gets in too big a hurry between *fast-forward*, *reverse*, and *play*.) The key to avoiding tape spillage or jamming with a cassette: Always remove the slack between reels before playing a tape or rewinding it, and always allow a few seconds for *play* before going from *reverse* to *fast forward* or vice versa. Before winding, playing or rewinding, tighten the tension on the takeup reel in the cassette by using a pencil or similar stylus to hand-wind a turn or two. These procedures serve to keep the tape taut between takeup and supply reels and prevent spillage and jamming.

The Popular 8-Track Cartridge

The cartridge contains a length of tape which is made to run in a continuous loop at a speed of 3¾ inches per second (ips). The tape travels across the front edge of the cartridge where the center opening permits contact with the tape head. A pressure pad and spring-tension assembly appears on the cartridge to maintain proper contact at all times between head and tape. The pinch roller, inside the cartridge, presses against the capstan (on the player); the tape, pinched between roller and capstan, is thus moved across the head.

There are eight separate track areas on the quarter-inch tape—one pair of stereo tracks for each of the four programs. Because the tape is a continuous loop, the beginning and end of all eight tracks are recorded adjacent to each other. A metal sensing strip is normally found at the end of each program; this actuates the track-shift solenoid coil and causes the tape head to step to the next program. The four different positions assumed by the tape head are shown in Fig. 1-2. At the end of program 4, the tape head will either automatically cycle to program 1 or shut off, depending on individual player design. A different program can be selected by pushing the track-shift button, which accomplishes manually what the metallic sensing strip does automatically.

THE 4-CHANNEL CARTRIDGE

The quadraphonic cartridge reproduces four-channel sound—a "different" sound from each speaker. On a stereo 8-track cartridge there are four programs; on a quadraphonic cartridge, there are but two. The good 8-track quadraphonic (Q8) cartridge divides the music into two four-track quadraphonic programs (Fig. 1-3).

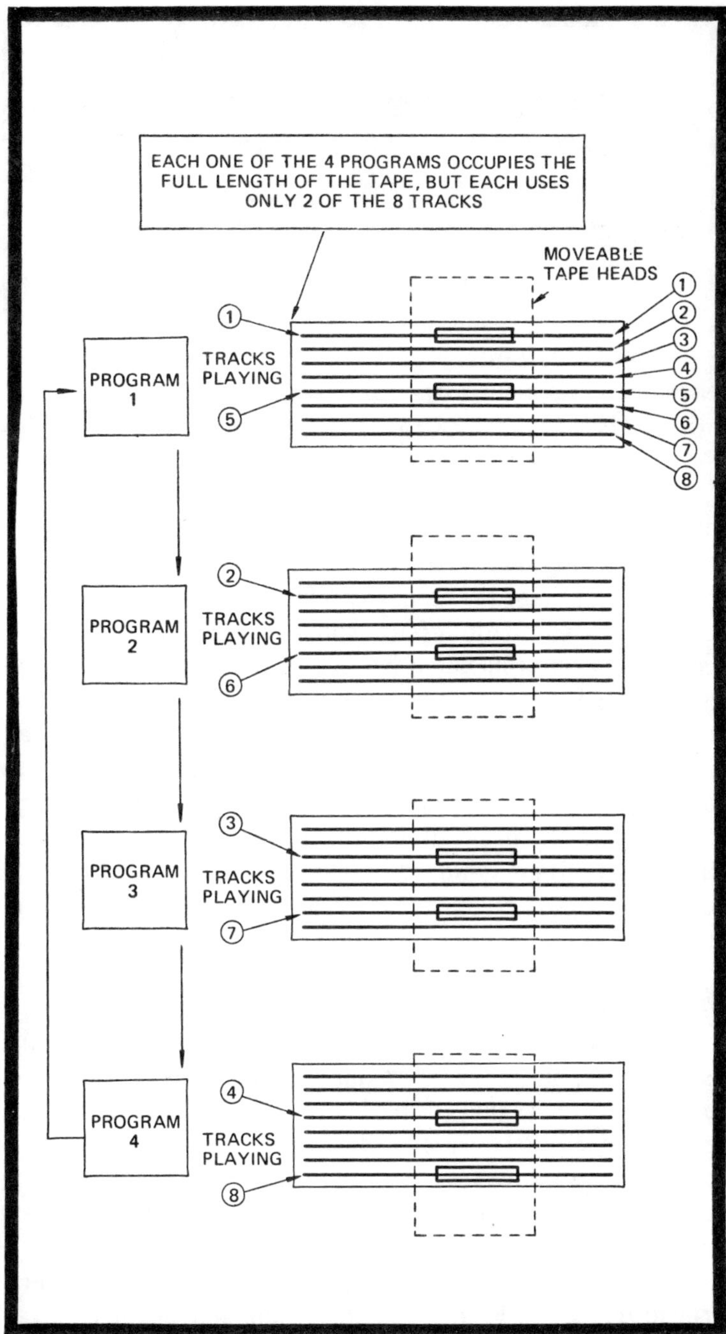

Fig. 1-2. The stereo 8 track and channel 6. (Courtesy Lafayette Radio.)

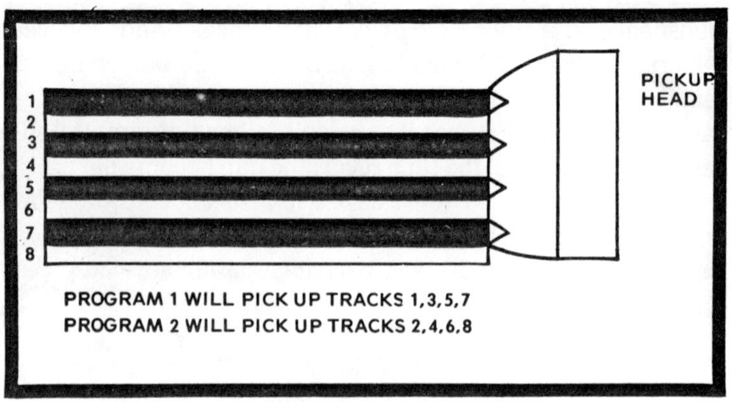

Fig. 1-3. Q8 tape track head pickup.

The Q8 cartridge is identical in size to the 8-track stereo cartridge and is fully compatible with the 8-track player. A small vertical notch in the top left corner identifies the four-channel cartridge to the player. When a Q8 cartridge is inserted in a compatible four-channel player, the tape head contacts the appropriate group of four tracks and plays to the end. At the end it automatically switches to pick up the second program. When an 8-track stereo cartridge is played in a compatible four-channel player, the tape head contacts only one pair of tracks at a time and automatically plays all four programs. Although 8-track stereo cartridges may be played on either standard or four-channel players, the Q8 cartridge should be played on a four-channel player to obtain proper effect.

SUMMARY OF TAPE CARE PRACTICES

Always remove tape cartridge or cassette from the player when not in use. Store tapes in a cool, clean, dry place. Tapes should be placed in plastic bags and stored in tape storage containers, free of dust, dirt, and other contaminants. Keep the tapes out of the sun, too; the tape case is made of high-impact plastic and will tend to deform if exposed to heat.

To avoid jamming, let cold tapes warm to normal temperature (65—70°F) before being played. Keep tapes away from strong magnetic fields (such as electric motors, loudspeakers, radios, electrical appliances, and TV receivers). In color TV models that have a tape player in the same cabinet, always remove and set tapes aside when degaussing the picture tube; if you don't follow this simple precautionary measure, the tape recording can become partially erased and damaged. Proper care will insure longer tape life and more satisfactory tape performance. And don't forget to remove those test and blank cartridges after servicing the tape player.

MECHANICAL CONFIGURATIONS

Even though there is an almost countless number of tape players on the market for use in cars and with home stereo systems, there are only a few basic configurations—and any tape player will fit into one of these categories: personal portable, stereo or Q8 music system, automotive accessory, and hi-fi system deck.

Cassette Players

The simplest popular cassette system is the "personal portable" model typically used for dictating messages and recording and playing back monaural musical selections (Fig. 1-4). Perhaps the most important cassette system is the type manufactured for use as a musical source in a high-fidelity stereo or four-channel music system (Fig. 1-5). Interestingly, the cassette fills both roles admirably, and a recorded stereo or quadraphonic cassette is completely compatible with the lowliest monaural portable player!

If you aren't familiar with tape players, the terminology and nomenclature of player components may complicate the issue. In this book, we'll be using the standard nomenclature of the industry, as exemplified by the list given below (and referenced to the exploded view shown in Fig. 1-6.

1. Mechanical chassis assembly
2. Guide spring
3. Guide spring shaft
4. Slide chassis assembly
5. Recording lock plate (A)
6. Recording lock plate (B)

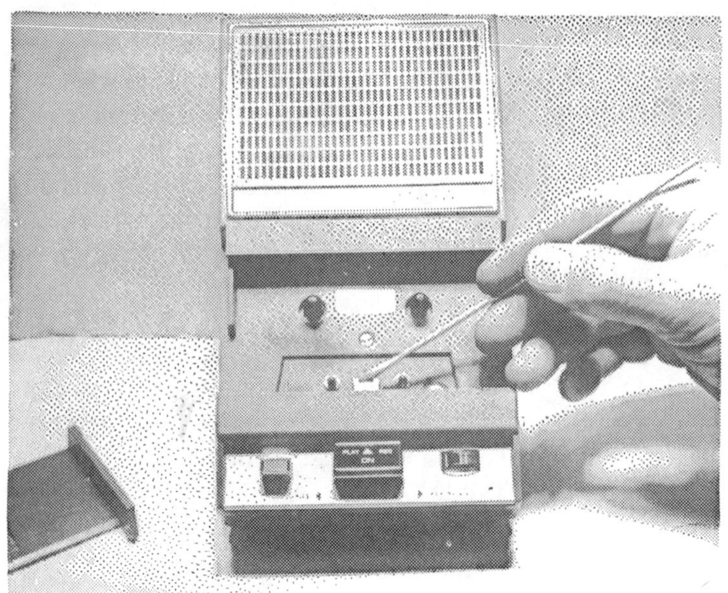

Fig. 1-4. A typical portable cassette tape player.

7. Recording lock plate wire
8. Amp supporter
9. Cassette holding plate support
10. Cassette holding plate support
11. Opening–shutting spring
12. Function lever arm guide
13. Friction lever arm
14. Rewind lever assembly
15. Slide roller
16. Pinch roller spring
17. FF rewind lever spring
18. Rewind lever spring
19. Head regulation spring
20. Cassette cover lock plate spring
21. Friction lever spring
22. Recording lock spring
23. Slide spring plate
27. FF rewind assembly
28. Poly washer
29. Guide roller

30. FF rewind slide plate assembly
31. Pinch roller assembly
32. Friction mechanism assembly
33. Reel base assembly
34. Reel base cap
35. Tension spring
36. Motor supporter
38. Capstan assembly
39. Capstan rubber belt
40. Motor pulley
41. Bracket
42. Top cover lock plate
44. Slide chassis spring
45. Recording lever assembly
47. Recording lever spring
48. Recording lever guide roller
51. Cassette-up lever support assembly
52. Volume support
68. Bottom body bracket
70. Slide chassis pressure spring
71. Volume knob
72. Pushbutton
73. Reel base washer
74. Slide chassis guide shaft
75. FF rewind lever shift
M—motor

Fig. 1-5. The cassette tape deck may have a record player, AM FM Radio, AM FM—FM Stereo, and large amplifier system in one unit. (Courtesy Allied-Radio Shack.)

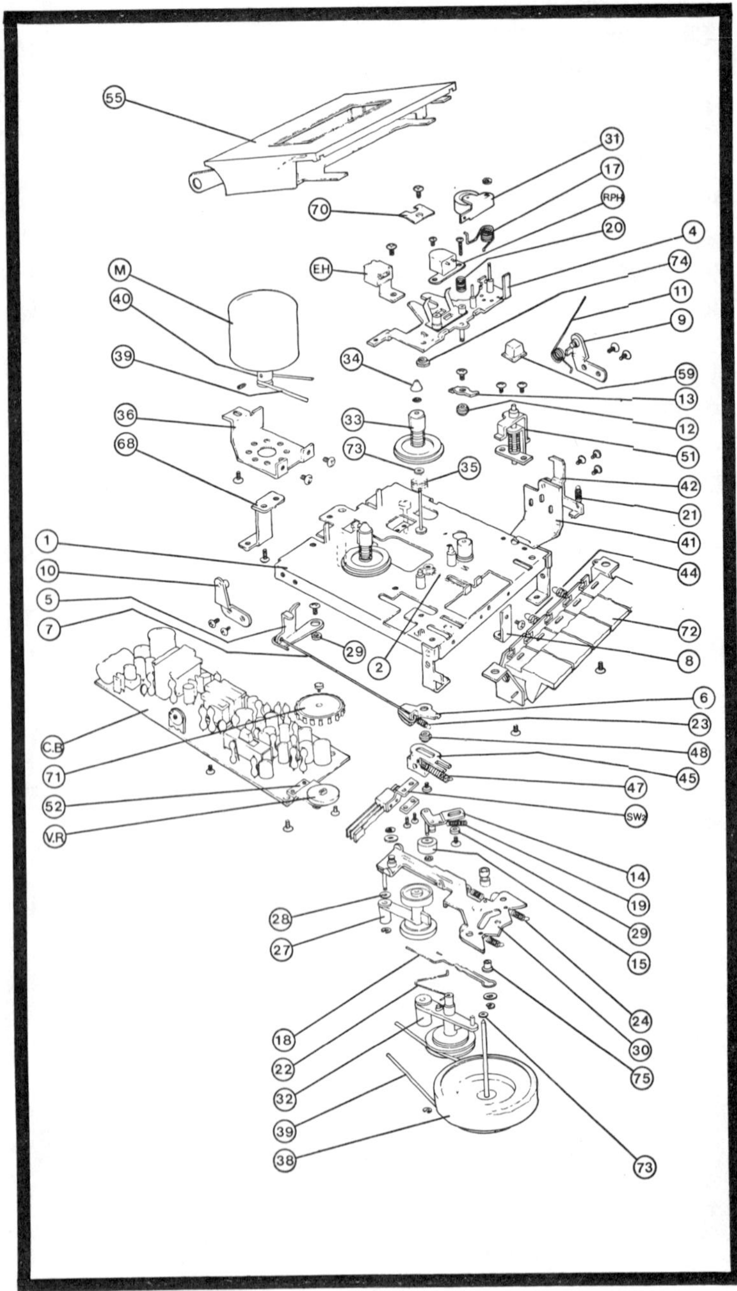

Fig. 1-6. Exploded view of cassette tape player. (Courtesy Automatic Radio Model CPR-2086.)

SW2 — reed switch
CB — circuit board
VR — volume (⅛?K ohm)
RPH — R–P head
EH — erase head

Cartridge Players

The 8-track car stereo tape player (Figs. 1-7 and 1-8) has by now become more or less standardized in design. The more recent Q8 players are similar in design; they will, of course, play both 2- and 4-channel tape cartridges.

Interestingly, the development of the cartridge player did not parallel the development of the cassette. The cartridge player found immediate acceptance for automotive installations, and the popularity of this system increased dramatically during the past 10 years or so. The improvements in design have been chiefly aimed at making the unit less expensive for the purchaser. As a result, cartridge player decks may sometimes be found with price tags of less than $20, even though a home-type unit can be purchased with

Fig. 1-7. A auto- stereo 8-track tape player shown with top cover removed.

Fig. 1-8. When the cover has been removed, auto 8-track units tend to look pretty much alike.

"quality sound" features deserving of an economy-minded audiophile's attention. The unit pictured in Fig. 1-9, for example, may be operated as either a 2- or 4-channel source and serves as the nucleus of a complete and fairly inexpensive sound system.

An exploded view of a portable 8-track tape player mechanism is shown in Fig. 1-10. The following parts list identifies each component:

1. Chassis
2. Ceiling w/pressure, cartridge
3. Bucket cam
4. Cam shaft clamp pen
5. Cam stopper
6. Cam shaft
7. PC board indicator
8. Channel selector assembly
9. Head height adjust arms
10. Adjust arm spring
11. Head arm screw

Fig. 1-9. The top cover has been removed from this combination AM-FM receiver and 8-track.

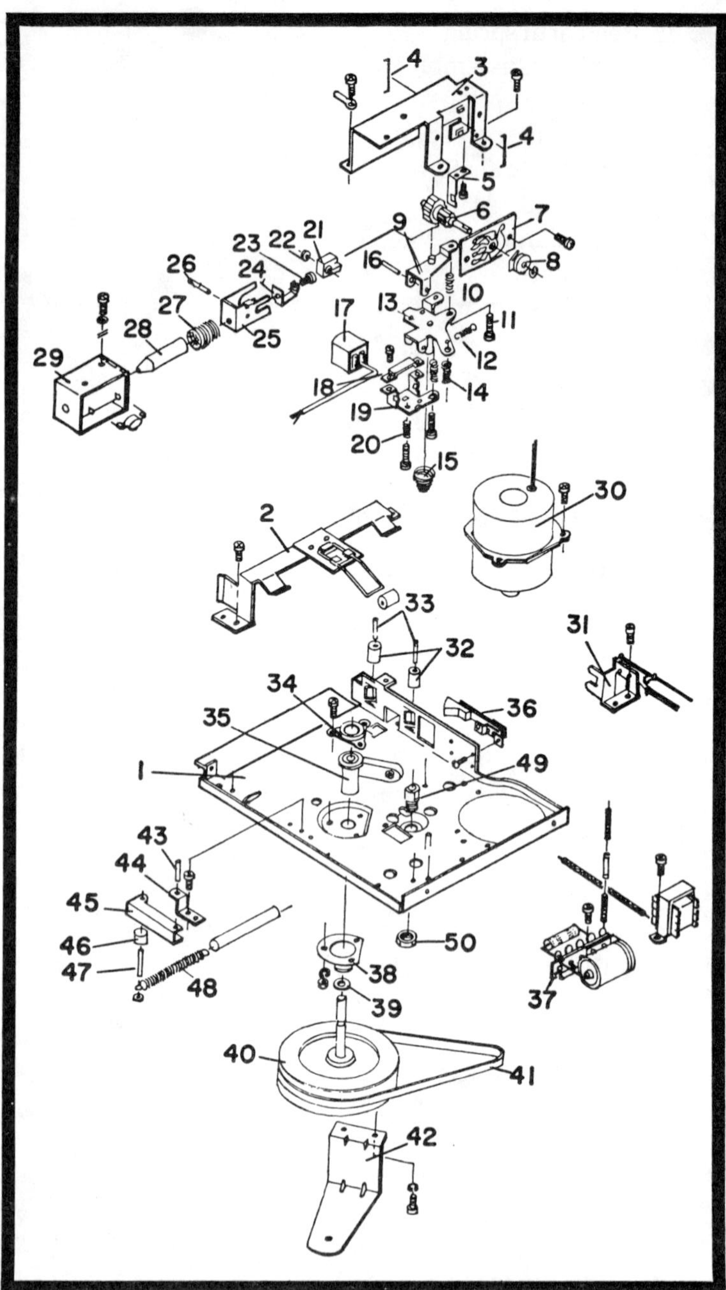

Fig. 1-10. Exploded view of stereo 8 tape portable player. (Courtesy Lloyds.)

12. Head arm spring
13. Head arm assembly
14. Head arm spring A
15. Head arm shaft spring B
16. Adjust arm shaft
17. Play head
18. Head clamp
19. Head base
20. Head base spring
21. Cam ratch
23. Plunger screw
24. Ratch pressure
25. Ratch bracket
26. Ratch shaft
27. Plunger spring
28. Plunger
29. Solenoid coil
30. Motor and pulley
31. Auto switch and tape guide assembly
32. Pressure roller
33. Pressure roller shaft
34. Flywheel disc plate
35. Bearing
36. Leaf switch
37. Terminal strip
38. Bearing holder
39. Flywheel nylon washer
40. Flywheel assembly
41. Motor belt
42. Flywheel support
43. Side arm holder shaft
44. Side arm holder
45. Side pressure arm
46. Side arm roller
47. Side roller shaft
48. Pressure arm (side) spring
49. Cam stud
50. Cam stud nut

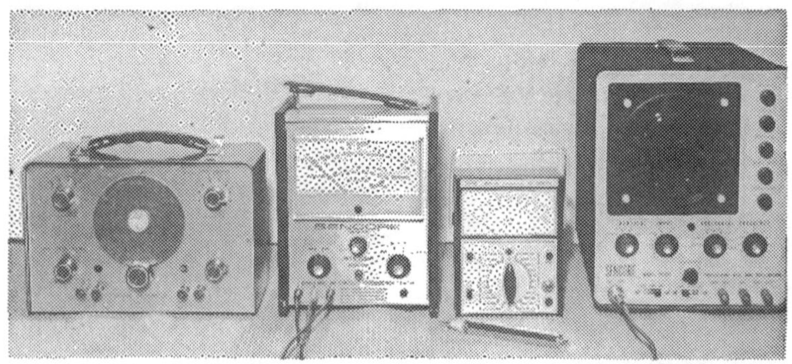

Fig. 1-11. Test equipment needed to service the cassette and cartridge tape players.

TEST EQUIPMENT FOR SERVICING

Only a few additional test instruments and tools are needed to properly service cassette and cartridge machines. Most of these test instruments are already found upon the electronic technician's service bench (Fig. 1-11). If you are already servicing solid-state car radios, auto tape player repair is a snap. Solid-state tape player audio circuits are no different from those found in automotive stereo receivers and amplifiers. Just add a few special test tapes and balance meters and you're in the tape player servicing business.

High-Impedance Voltmeter

The VTVM, FET-VOM, or other high-impedance (not less than 100K per volt) meter can be used for at least the following:

1. Measurement of small transistor voltages
2. Height adjustment indicator
3. Azimuth adjustment indicator
4. Continuity checks
5. Checking motor voltage
6. Measurement of record oscillator bias voltage.
7. Measurement of erase voltage
8. Playback amplifier sensitivity indicator
9. Record amplifier sensitivity indicator
10. Checking of ALC (auto level control) characteristics
11. Checking for bias voltage on tape head

Audio Signal Generator

The audio signal generator is used in the following procedures:

1. Checking tape head alignment.
2. Signal tracing defective audio circuits.
3. Signal source for determining weak audio stages.
4. Signal for checking reproducing amplifier sensitivity.
5. Signal for checking record amplifier sensitivity.
6. Measurement and connection of frequency response characteristics.

Oscilloscope

The scope is used for the following service procedures:

1. Audio signal tracing.
2. Locating and isolating distortion in audio stages.
3. Indicator for height and azimuth adjustments.
4. Making record bias frequency adjustments.
5. Checking record bias patterns.
6. Checking presence of bias voltage on tape heads.

Universal DC Power Supply

If you're planning on taking tape player repair seriously, you'll need a dc power supply to furnish operating power to the auto stereo tape player while playing or making adjustments on the service bench. It is best to choose a variable supply to handle the largest car radios and stereo tape players. Some power supplies have an integral voltmeter and ammeter to show the applied voltage and current drawn by the tape player. A power supply suited for tape player repair should have a variable voltage range up to 16.8V and maximum 10-ampere constant current rating. The variable voltage range will help locate an intermittent audio transistor. When the voltage sharply drops and excessive current is measured, a direct short or overloaded circuit is indicated.

Test Tapes

Several 8-track standard stereo test tapes are needed to simplify checks for crosstalk, azimuth, and height adjustments, frequency response, and stereo separation. A blank test tape should be used when turning the tape player over and around in making service tests on an 8-track stereo tape player. Generally, only one test tape covering the above specifications is needed to check out the cassette player. A special torque test cassette is handy when checking the cassette tape player takeup reel torque and locating a defective clutch assembly.

Miscellaneous Test Equipment

A pencil-type noise generator can quickly locate a dead or weak audio stage by moving it from base to collector of each transistor. The audio signal tracer is handy to locate a very weak or distorted stage in the audio circuitry. A suspected defective tape head can be checked by simply clipping the audio signal tracer to the tape head wiring. The direct output voltage of the recording tape head can be tested with an audio VTVM, although they are expensive. VU meters can be used to check audio balance and tape head adjustments. A tension test gage is used to make pressure adjustments (in grams).

Amplifier and Power Circuits

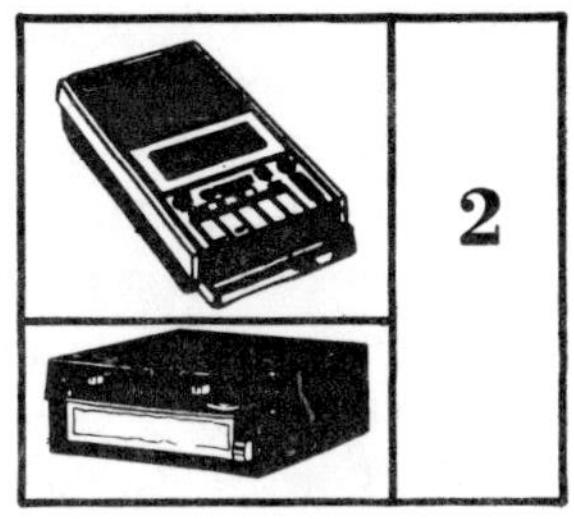
2

Audio amplifier problems are somewhat more common with automotive stereo equipment than with the power-line operated equivalent units. This may be partially attributable to the fact that the automobile environment is more severe (in terms of temperature extremes, vibration, shock, dust and dirt, and similar conditions); but perhaps even more important is the wide range of applied-voltage conditions the circuits must withstand. A mobile tape player installation receives supply voltages that may vary from as low as 10.6 volts under no-charge, heavy-drain conditions to as much as 16 volts. By way of comparison, that input-voltage spread represents a variation equivalent to a power-line range of 106 to 160 volts!

With a voltage range of such magnitude, it is not surprising to see the anomalies that tend to occur in tape player circuits. Excessive voltages applied to resistors cause overheating, which might in turn cause the resistors to change value permanently. Capacitors, too, change value as a result of overvoltage and overheating. An electrolytic, in particular, has a tendency to change value with a change in voltage. And when an electrolytic has been used in-circuit for an extended period at some voltage other than its design value, there may be very little similarity between that capacitor's *marked* value and its *actual* value. This problem can be quite bothersome with transistor circuits—for some very obvious reasons. For openers, consider the fact that transistor circuits in general have low input and output impedances. A low impedance means an uncommonly high value of capacitance in the coupling circuit. What happens when a high-value capacitor becomes a low-value capacitor? If the capacitor is used to couple an audio circuit from one stage to another, it means the lower ranges of audio frequencies to be passed through the capacitor will be attenuated.

Consider for a moment the effect of a changed-value capacitor in an amplifier's emitter bypass. Such problems were nonexistent during the days of the vacuum tube. Impedances were high and capacitance values low—even in cathode bypass circuits. (An electrolytic as a bypass was unheard of with vacuum-tube amplifiers—with transistor amplifiers, they're the rule rather than the exception.)

If you recognize that electrolytic capacitors are imperfect devices and they tend to adjust themselves in value according to the environment in which they are used, you're well on the road to being expert in repair of solid-state audio devices. Bear in mind always that a capacitor is in itself a very basic series high-pass filter, and that an upward shift of value (higher capacitance) means lower frequencies can be passed through the capacitor with less attenuation. And remember that a downward shift in value (less capacitance) results in low-frequency loss. The capacitor values actually meant for the circuit were chosen as an optimum compromise between wideband frequency response, impedance matching, effects on circuit stability, etc. When a customer comes to you with frequency-response problems that can't be solved with a little bottle of tape-head cleaner or head-alignment touchup, he could be having problems with shifting component values. Not that it is always easy to check. It's easy enough to measure the value of a resistor, but not until the advent of sophisticated ICs could we measure capacitor values in the same direct way in the field. And even today, the capacitance-measuring VOM is priced at well over $100.

Transistor amplifier problems are also complicated with direct-coupled audio stages. What happens when one stage has a bias resistor whose value has shifted? The net effect is a bias shift that affects all stages in that amplifier—particularly where the faulty stage is not in the preamp but several stages down the line in the circuit. Here, the customer will complain that the amplifier goes into distortion without attaining full volume. If you observe such an amplifier's signal on the scope, you'll note a single-ended class A amplifier may actually be operating class B! It is extremely important to keep each stage biased so that its quiescent-state operating point is precisely in the center of the linear portion of that device's transfer characteristic curve. Maximum volume occurs at that point and that point only. More bias or less bias will cause clipping of the signal (and distortion) at high stage gains.

PREAMP AUDIO CIRCUITS

High-impedance circuits are more susceptible to the effects of stray fields, oscillations, and the like. It will help immensely to remember ways of spotting high-impedance circuits when you're troubleshooting amplifiers plagued with hum, electromagnetic interference, and other such problems. Look for a high value of capacitance in the coupling circuit along with relatively low resistor value from the stage's input to ground.

As a matter of fact, you can closely estimate the circuit's lower cutoff frequency by calculating the RC circuit's time constant and converting that to frequency by the reciprocal method. (Multiply capacitor value by resistor value, then divide 1.0 by the product.) In Fig. 2-1 the playback head is coupled through a 20 μF capacitor into a load whose resistance totals about 50K. Just convert microfarads to farads (20 μF equals 0.00002F), and multiply by 50,000. As can be seen, the answer is 1.0 (second), which equates to a signal frequency of 1 Hz, well below the design frequency cutoff of the head itself.

In Fig. 2-2 the coupling capacitor is 1 μF and the load impedance is 39K; so the period of the circuit is 0.000001 × 39,000 or 0.039. This period translates to a low-frequency cutoff of 1/0.039 or approximately 25 Hz—still better than the low-frequency limit of the tape head itself.

To locate problems in the preamp circuits, take voltage measurements, apply signal injection and signal tracing. In stereo preamp circuits the signal in one channel can be traced and compared with the signal in the other audio channel at its comparable position in the overall chain. Always keep the audio signal low in level when injecting test signals into the preamp stages. To quickly isolate an audio problem, apply the test signal to one of the stages between the volume

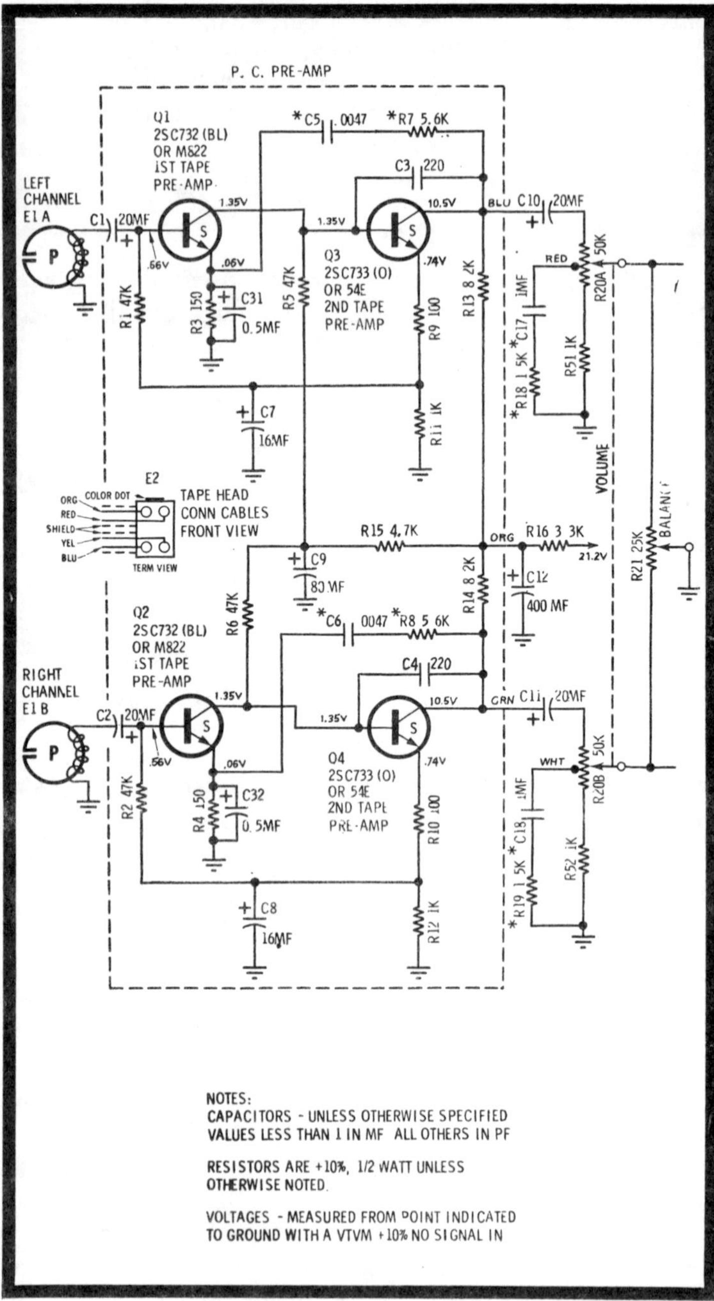

Fig. 2-1. Preamp direct-coupled circuit of a stereo 8-track tape player. (Courtesy of Masterwork.)

control and the tape head. You can inject signal from collector to base of each audio transistor until a loss of signal is heard. After the sound problem has been traced to the faulty preamp stage, it is a simple matter to ascertain the defect within that stage.

Getting to the root of a problem that has been isolated to a particular stage depends on how well you know the function of the component in that stage. When the active device (transistor) in the stage is known to be good, locating the bad part is pretty much a cookbook procedure. If *audio is distorted and excessive,* look for an open or disconnected emitter bypass capacitor. If it isn't open or disconnected, start suspecting a significant upward shift in value. If *audio is undistorted but weak,* don't overlook the possibility of a downward value shift here—but be careful to avoid introducing distortion when you replace the emitter bypass, because this capacitor is intended to reduce the stage's gain somewhat by virtue of the negative feedback characteristic; it introduces a portion of the stage's signal back into the stage 180 degrees out of phase with the input in order that all distortion introduced by the stage itself can be canceled completely.

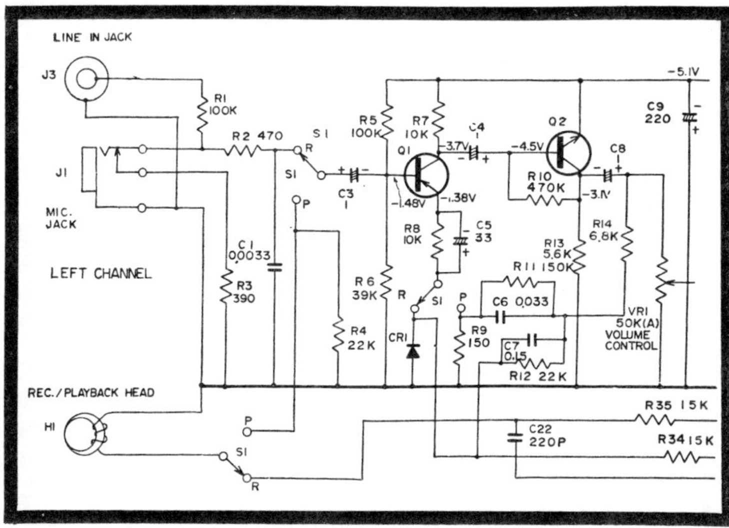

Fig. 2-2. Resistance-capacitance circuit in preamp stages of a portable cassette player. (Courtesy of Allied/Radio Shack.)

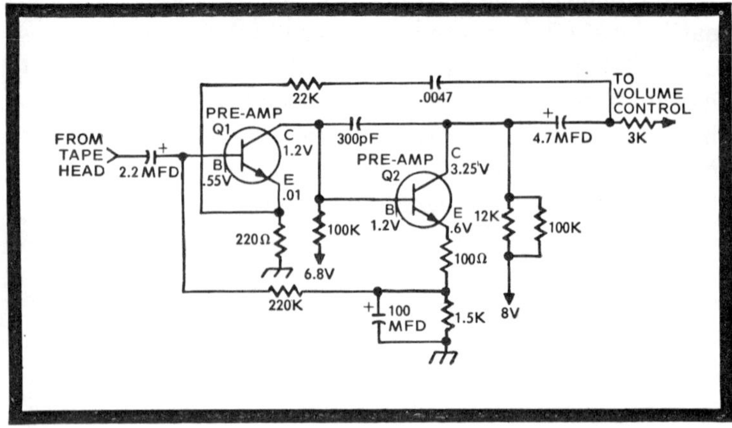

Fig. 2-3. Preamp direct-coupled circuit of a auto stereo 8 tape player with voltage measurements.

If the audio out of the stage is below the level of the equivalent stage in the opposite channel, and this signal loss is accompanied by hum, an impedance mismatch is indicated. Check the values of the series coupling capacitor on the input, but be especially on the lookout for a disconnected resistor from the input to ground. If the resistor is not disconnected, suspect an increase in resistor value. Bear in mind that optimum transfer of energy occurs when the impedance of the load is the same as the impedance of the source. If the input of that stage is higher resistance than the source resistance, not only will the audio signal actually decrease but the input circuit will be prey to whatever stray fields happen to be in that vicinity of the amplifier.

Transistor Defects

Voltage measurements taken at the preamp can usually serve to quickly locate a defective transistor in-circuit. It is always preferable to use a VTVM or FET-VOM to make these low-voltage readings, although some technicians do get by handsomely using ordinary VOMs with high input resistances (100K or so). A typical auto-stereo preamp circuit is shown in Fig. 2-3, with normal operating voltages given adjacent to the terminals of each transistor. When either transistor becomes open or leaky, a voltage change will be noted on each transistor.

By following the voltage change chart of Fig. 2-4, we can easily locate the defective transistor in the two-stage circuit of Fig. 2-3. Let us assume, for a moment, that transistor Q1 has an internal open circuit. This condition will be indicated by an increase in emitter voltage (on both transistors), a near doubling of Q2 base voltage, and a drastic reduction in the collector voltage of Q2. With a leaky Q1:

1. The base and collector voltage of Q1 and base of Q2 are the same.
2. The collector voltage on Q2 is high—actually equal to supply voltage of that circuit.

With an open condition in Q2:

1. No or low voltages on base and emitter of Q1.
2. Collector on Q1 rises to supply voltage for that circuit.
3. Voltage on base, emitter, and collector of Q2 are high.

With a leaky condition in Q2:

1. Voltages on Q1 may be normal or nearly normal.
2. Base and emitter voltages may be near normal.
3. Q2 collector voltage drops by half.

In case either transistor has a direct internal short, the voltage may be the same on all terminals of the transistor and much lower than the supply voltage. Generally, when a

TRANSISTOR	Q1			Q2		
	E	B	C	E	B	C
NORMAL VOLTAGES	.01V	.55V	1.2V	.6V	1.2V	3.25V
Q1-OPEN	.575V	.9V	1.5V	.95V	1.5V	1V
Q1-LEAKY	.01V	.55V	.55V	.05V	.55V	8V
Q2-OPEN	0V	0V	7V	6.5V	7V	8V
Q2-LEAKY	.01V	.575V	1.25V	.75V	1.25V	1.75V

Fig. 2-4. Transistor voltage chart of actual measurements with leaky and open conditions.

transistor opens, the collector voltage may be near the supply voltage. In direct-coupled circuits, voltage readings on one transistor will more drastically affect voltage readings upon the other.

Since transistors themselves represent some 90% of the defects in solid-state circuitry, some service technicians prefer to first check the condition of each transistor. The transistor can be checked accurately in the circuit for open conditions with a transistor beta tester.

For accurate beta and leakage readings, remove the collector terminal of each transistor from the circuit. When leakage tests are made on direct-coupled circuits with an in-circuit transistor tester, a high leakage reading is noted. When in doubt, remove the suspect transistor from the circuit.

Integrated-Circuit Defects

In the last two years ICs have found their way into many cassette and cartridge tape players. At first, the IC chip was used only in the audio preamp circuits. Today, though, in some auto stereo and home tape players, the entire audio system may be incorporated into a couple of integrated circuits. In most cases the preamp circuitry consists of one IC chip; and the driver and output power circuits are combined into another chip.

The IC chip may consist of many transistors, resistors, and (sometimes) capacitors. Since space is limited inside the chip area, the resistors are quite small in terms of power handling ability. Similarly, the values of IC resistors tend to be on the low side—which means larger-than-normal external capacitor values for any given circuit impedance.

On most circuit diagrams, the integral transistors and resistors of an IC are not shown. Instead, the IC will be depicted by a triangular symbol, as in Fig. 2-5. Notice the triangle "points" toward the audio signal path. You will notice in the stereo tape player preamp there are very few components tied to the IC. the external components include only the compensation circuitry and those capacitor values that could not be incorporated into the chip because of physical size constraints. In these cases, the IC becomes a "black box" and troubleshooting is limited to that portion of the audio circuit that is external to the chip. IC problems are dealt with by the expeditious practice of replacement.

When drawn within a circuit diagram, the IC appears as a triangle. But don't make the mistake of looking for a device on a board with this configuration. In practice the IC chip's physical shape may be round, rectangular, or square. Usually there will be between 7 and 14 leads; but some tape units have been known to employ three-lead devices! The manufacturer ordinarily numbers each leg for correct circuit connection. Always check for an indentation on the case of a symmetrical-lead IC before removing the device from a circuit. In most cases these ICs can be turned around and inserted backwards. The indentation (Fig. 2-6) allows you to correctly orient the device before you reinstall it; putting it in backwards can mean quick destruction of a fairly expensive component.

In tape player audio circuits a defective IC can be located with signal tracing. Inject an audio signal at the player's volume control to determine if audio output circuits are functioning; a pencil-type audio generator works nicely here. If the audio tone is good, proceed to the output terminal of the IC preamp circuit. Then go to the input terminal of the IC. In case of no or weak volume the preamp IC circuit is defective.

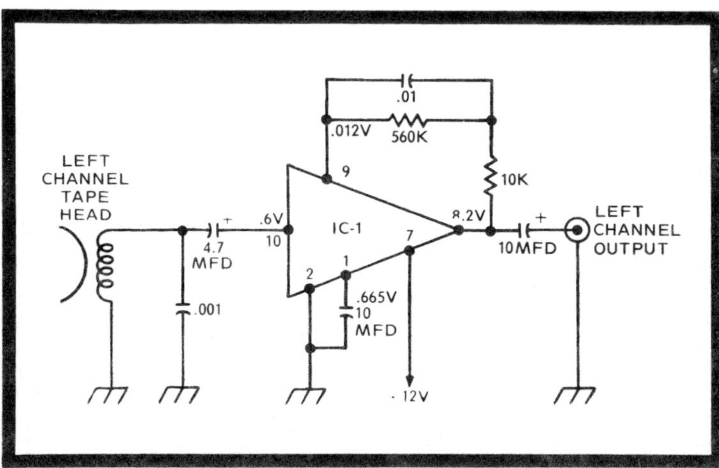

Fig. 2-5. IC preamp circuit of a stereo 8 home player that plugs into stereo AM-FM/phonograph combination.

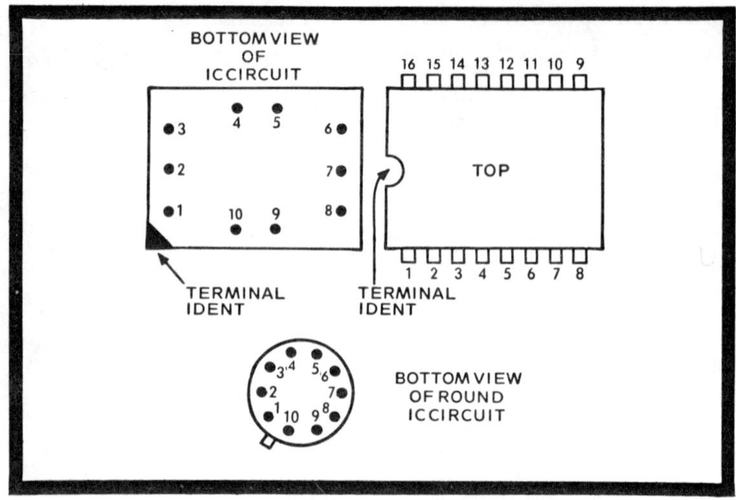

Fig. 2-6. Terminal connections on IC circuits found in stereo 8 amps.

To assure yourself that your diagnosis of a bad IC is correct, take voltage measurements. Compare your actual voltage measurements with those values shown on the schematic. (If a schematic diagram is not available, compare the voltage measurements with the good IC chip in the other stereo channel.) Be careful not to short the voltage test probe against an adjacent IC terminal, or the bad IC may be the result of your own carelessness.

You can replace the IC to determine if the circuit is defective, but this is normally considered "bad form"—especially when the IC is soldered in place. Also, remember that there are many of these IC audio circuits in use today and it is quite impractical for you to maintain a stock of them all. In case you service a couple of different brands or specialize on warranty of these units, then you can have several IC replacements on hand. But still, replacement of the suspect IC may mean unsoldering up to 14 (sometimes even more) circuit board terminals. And often, an IC that might otherwise have been good will be destroyed by the applied heat during removal from the circuit. When replacement is unavoidable, the best method is to use a suction iron that removes excessive solder from the connection. All connections must be unsoldered before the IC chip can be removed from circuit board.

CASSETTE RECORD–PLAYBACK HEADS

In most cassette players you'll only find a single head (other than the erase head). In the unit pictured in Fig. 2-7, the tip of the ballpoint is touching the record–playback head; the white head to the left is for erasing. The recording circuit's requirements are somewhat different from those of the playback circuits. The ideal head for one is not ideal for the other; one requires a sharp, narrow gap while the other needs one slightly wider with a different gap angle. A single-head unit offers a reasonable compromise by offering electronic compensation for the compromise's shortcomings in either mode (playback or recording). Since the penalties do not involve degraded audio reproduction capability, the single-head approach is the standard, even on most expensive "audiophile" record–playback decks.

A defective tape head may produce *no output, weak signal, intermittent sound, crosstalk, or distortion.* Weak or distorted audio will most often be caused by a tape head that is encrusted with tape-oxide contaminants—in this case, of course, a thorough tape head "cleanup" may solve the problem completely. Generally, intermittent sound traced to

Fig. 2-7. Portable cassette player showing erase and R/P tape heads.

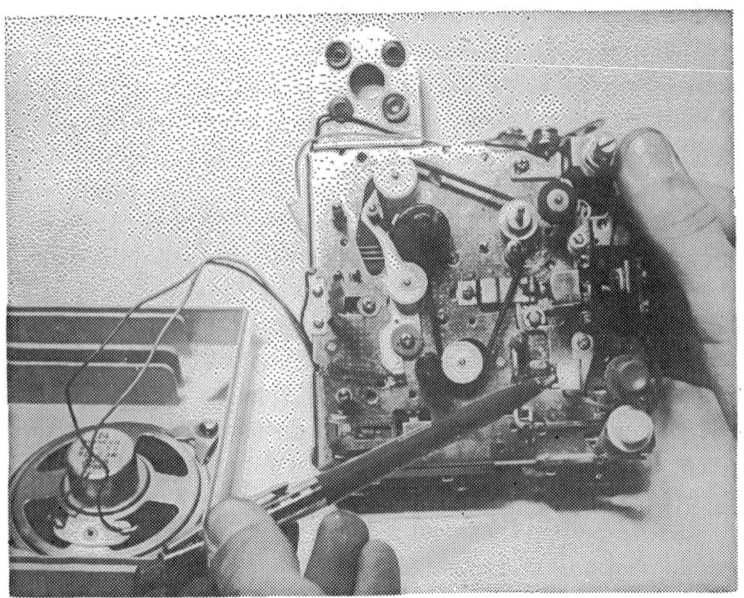

Fig. 2-8. Portable cassette player with magnet as erase head.

the tape head is caused by defective shielded cables. While these cables will seldom cut through and ground to the shielded cable, they are subject to moving and flexing every time the player is switched from playback to off and vice versa. Check the solder connections at the rear of the head by jiggling the wires soldered there carefully. (Use a long, fine tool for this—preferably a fiber flat-blade tuning tool or the like.) A loose tape head input plug can also produce intermittent sound. Distorted sound may be traced to a fouled head gap. Cross modulation between channels, or two channels with different programs, will most likely be caused by improper head alignment, although poor-quality tape is a notorious troublemaker in this respect.

The cassette tape head circuit can be checked with an ohmmeter; depending on the amplifier circuit for which the head was designed, the resistance of the tape head should match the input impedance of the first audio stage—usually between 150 and 600 ohms. Generally, the cassette tape head resistance will be the same on many different cassette players of the same manufacturer. You can check the tape head by substitution. Replacement of the tape head as a "substitution" check is normally impractical because of the problems involved with stocking various cassette tape heads. The tape head can be signal-traced, while in operation, with another amplifier connected to the tape head terminals.

CASSETTE ERASE HEADS

The erase head of a single battery-operated cassette player might consist of nothing more than a magnet affixed to a movable arm. This magnet (Fig. 2-8) is positioned againstthe tape during recording. In more sophisticated units, the erase head is more conventional. A cassette erase head may be energized by a *bias oscillator* circuit. The resistance across an erase head driven with a bias oscillator circuit may be as low as an ohm. The resistance across the erase head in a dc circuit is generally above 400 ohms.

Besides supplying an erase voltage, the bias supplies an ac bias to the recording head. To improve signal fidelity the bias frequency used is commonly between 50 and 100 kHz. Distortion may be noted when the bias oscillator voltage is absent from the record head. The erase head of a small cassette tape player may operate directly from a dc voltage, as shown in Fig. 2-9. In better players, however, the bias oscillator feeds a signal to the erase head as well as the record head (Fig. 2-10).

The oscillator circuit may use one transistor or two transistors in a push-pull circuit (Fig. 2-11). The cen-

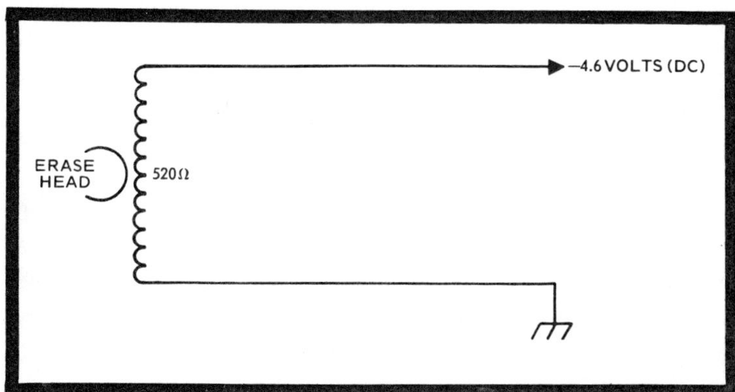

Fig. 2-9. Erase head circuit with a dc voltage applied directly to erase head terminals.

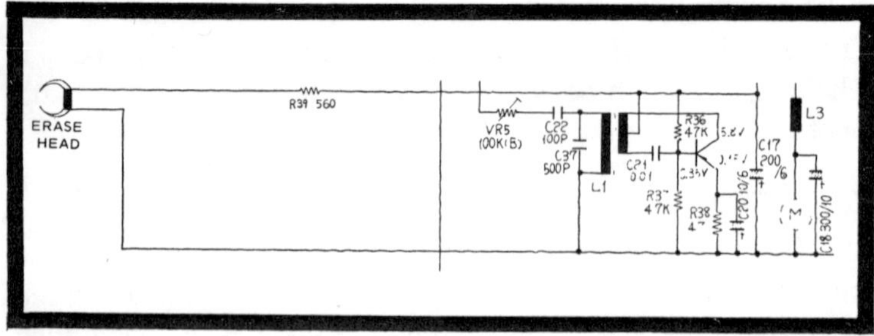

Fig. 2-10. Simple bias-oscillator circuit feeding to both erase and R/P heads.

tertapped transformer is constructed like a small i-f transformer and can be tuned from the top side. The push-pull oscillator circuit provides a greater current to the erase head, and is found in some of the more expensive machines that have an abundance of space.

The output of the bias – erase oscillator is often fed to the recording head through an RC network. In most bias circuits, a bias control (not shown in Fig. 2-11) is used to obtain the right amount of bias for minimum noise and distortion. When a recording sounds mushy and distorted, suspect a defective bias stage. Excessive noise and improper erasure of the tape also can be traced to this. The oscillator output signal is a sine wave and can be observed with an oscilloscope (Fig. 2-12).

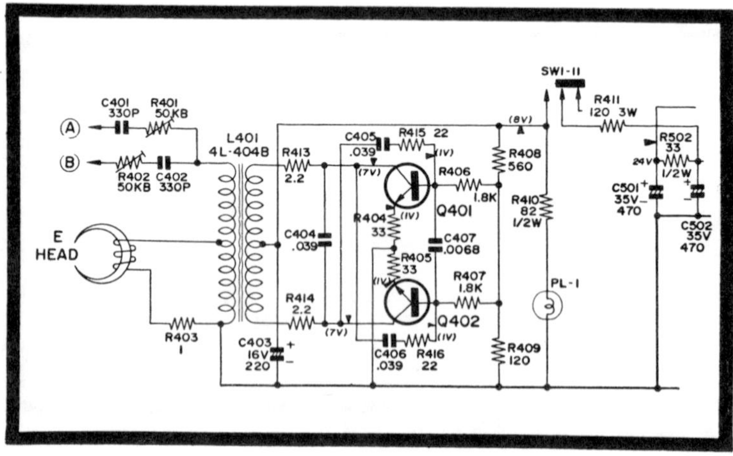

Fig. 2-11. Push-pull oscillator bias circuit.

CASSETTE AUDIO CIRCUITS

A simple cassette audio circuit may consist of a preamp, a driver, and a push-pull output stage. A larger cassette portable may have another driver stage as well as the oscillator—bias (and automatic level control) circuit. The audio circuit functions in both record and playback operations.

The preamp stages are usually *direct-* or *RC-coupled* transistor circuits. Generally, the audio driver stages are *direct-coupled.* An input driver transformer may be found between driver and output circuits. The push-pull output circuit is *transformer-coupled* to the speaker. A basic play-position audio circuit is shown in Fig. 2-13.

Push-Pull Output Circuits

Problems occurring in the push-pull output circuits include absent, weak, intermittent, or distorted audio. Again, assuming the tape heads are clean, transistors account for 90% of audio failures. The audio output stages can be signal-traced by injecting an audio signal at the speaker and proceeding toward the volume control. Voltage measurements

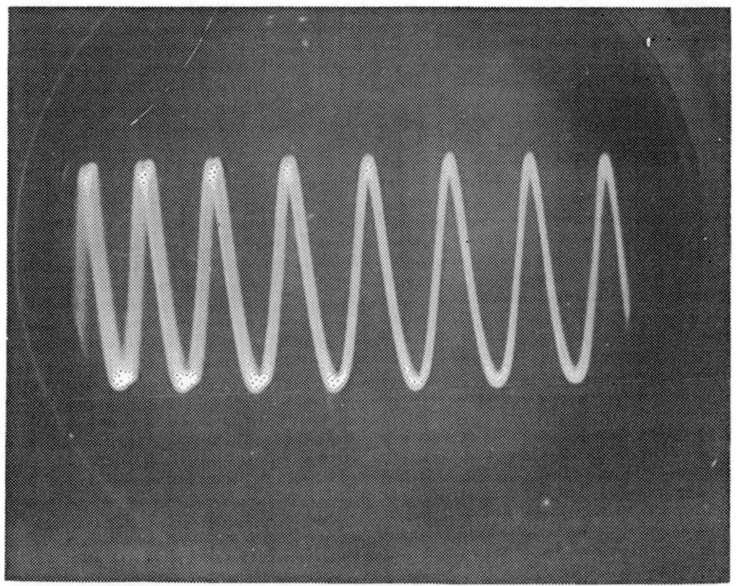

Fig. 2-12. A scope waveform taken at the recording tape head.

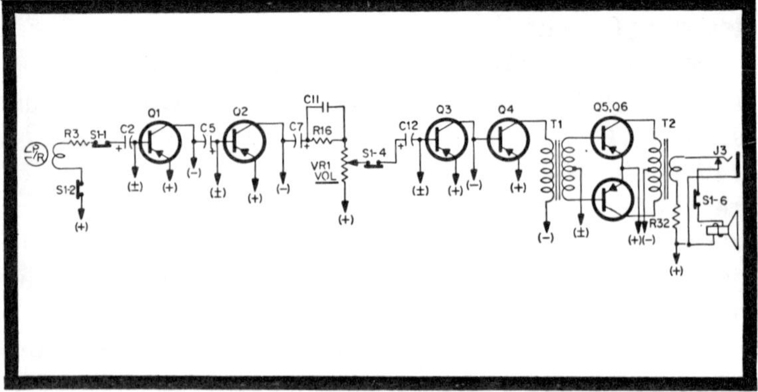

Fig. 2-13. Signal path—play position.

will usually uncover a defective output transistor immediately.

Weak and distorted sound can be caused by a leaky or shorted output transistor. Check the audio circuits for a burned bias resistor. Use your sense of smell to support your eyeballing operation here. A burnt or broken resistor may be hard to see and you may find that you need a magnifying light to locate it. Check the resistance of the transformer windings. An open primary winding of the transformer that connects to the driver collector (in Fig. 2-13) will produce no audio signal.

A defective output transistor often causes distortion. In many cases, the transistor will open; this condition can be checked in-circuit with a transistor tester designed for leakage tests. If a removed push-pull transistor appears normal, the other push-pull transistor can be checked accurately in the circuit for an open or leakage condition. (When one transistor is burned out in a push-pull stage, the associated transistor is often shorted or leaky.)

The base and emitter resistors have a very low resistance and should be checked on a low ohmmeter scale. It's best to remove one end of the resistor for accurate measurement. After a transistor has been pulled from the circuit, take the time to measure base and emitter resistances. Check the circuit diagram for correct resistor values; or, if a schematic is not available, compare the circuit with a similar output circuit of the same manufacturer.

Recording Circuits

For recording, ALC and bias oscillator circuits are added to the playback circuitry. A simplified record-position circuit is shown in Fig. 2-14. The automatic level control (ALC) stage is similar to an audio "compressor"—it increases the gain of the recording circuits when the input signal drops and reduces it when the signal is excessive. Thus, as the recording signal varies, the audio at the recording head remains relatively constant.

When the cassette is switched to the record mode, the microphone is switched into the preamp circuit and the audio output circuit is switched into the record head circuit. At the same time, the bias oscillator is switched into operation and fed to the record and erase heads.

Because of "shared" circuitry, audio problems may appear during both playback and recording. For example, if playback is weak or distorted with a prerecorded cassette, the same condition will likely exist in the record mode. If weakness or distortion is noted only in the record position, check the bias oscillator. The erase head circuit may also be affected, since the bias oscillator signal is also fed to the erase circuit. Crosstalk, or noise from a previous recording, will appear on a new recording if the erase head circuit does not function properly. Check for a bias sine wave on the re-

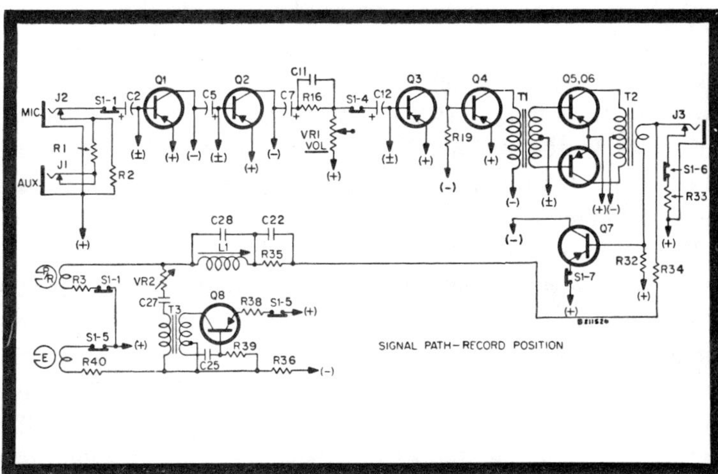

Fig. 2-14. Signal path—record position. (RCA Model YZB520.)

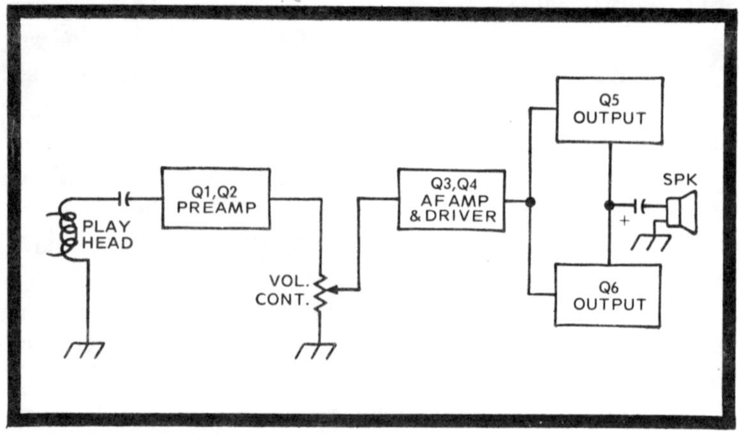

Fig. 2-15. Block diagram of typical channel of stereo audio stages.

cord head. An improper amount of head bias—or no bias—may be found. Check the operation of the record switch to see if it is engaging, and look for poor connections on the board.

8-TRACK AUDIO CIRCUITS

Sound problems in stereo tape player circuits are not much different than those in any audio equipment. In most cases, you will find one channel dead, or its output weak or distorted. A big advantage in servicing stereo circuits is the possibility of comparing voltages and signal levels in the malfunctioning channel with "good" sources in the properly functioning channel. When a component is not properly marked or is burned beyond recognition, you can locate it in the other channel for identification. With transistors, you can identify whether a transistor is a *pnp* or *npn* and find out its circuit application and proper operating voltages. When both channels are inoperative, suspect trouble in the power supply.

The typical 8-track audio section consists of a preamp, a driver, and a push-pull output stage (Fig. 2-15). The preamp stages are similar to the cassette preamp stages; service them as explained earlier. The driver and push-pull output stages sometimes are similar to those in low-output, inexpensive players. Generally, however, another audio stage and a transformerless push-pull output circuit are found in 8-track tape players (Fig. 2-16).

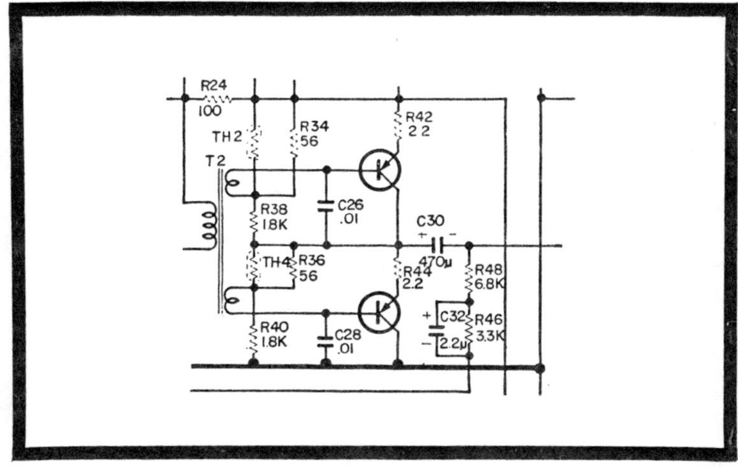

Fig. 2-16. Transformerless audio output circuit. (Courtesy Lloyds Inc.)

Generally, the audio output stage is direct-coupled to the af (audio-frequency) driver. In most cases, these stages are of the *npn* type. You will find very low voltages upon these stages. The driver stage is transformer-coupled to the push-pull output stages. The driver transformer is often physically larger and has more audio driving power than those found in the small cassette players. The secondary winding of the driver transformer is connected directly to the base of each output transistor. Complementary-symmetry output transistors will always be *pnp* and *npn*, as shown in Fig. 2-17. The speakers are coupled by capacitors to the output stages in Fig. 2-17. These are typically electrolytic capacitors of up to 1000 μF.

Problems sometimes found in the transformerless push-pull output stages include *no output* and *weak or distorted* audio. When one output transistor shorts or opens up, it will blow the main fuse; of course, everything goes dead. You can easily spot these defective stages by their burned bias resistors and overheated output transistors. Remove the suspect transistor and test it in a transistor tester. While the transistor is out of the circuit, check the other output transistor.

An intermittent or dead audio stage can be caused by an open condition or a poor connection of the electrolytic

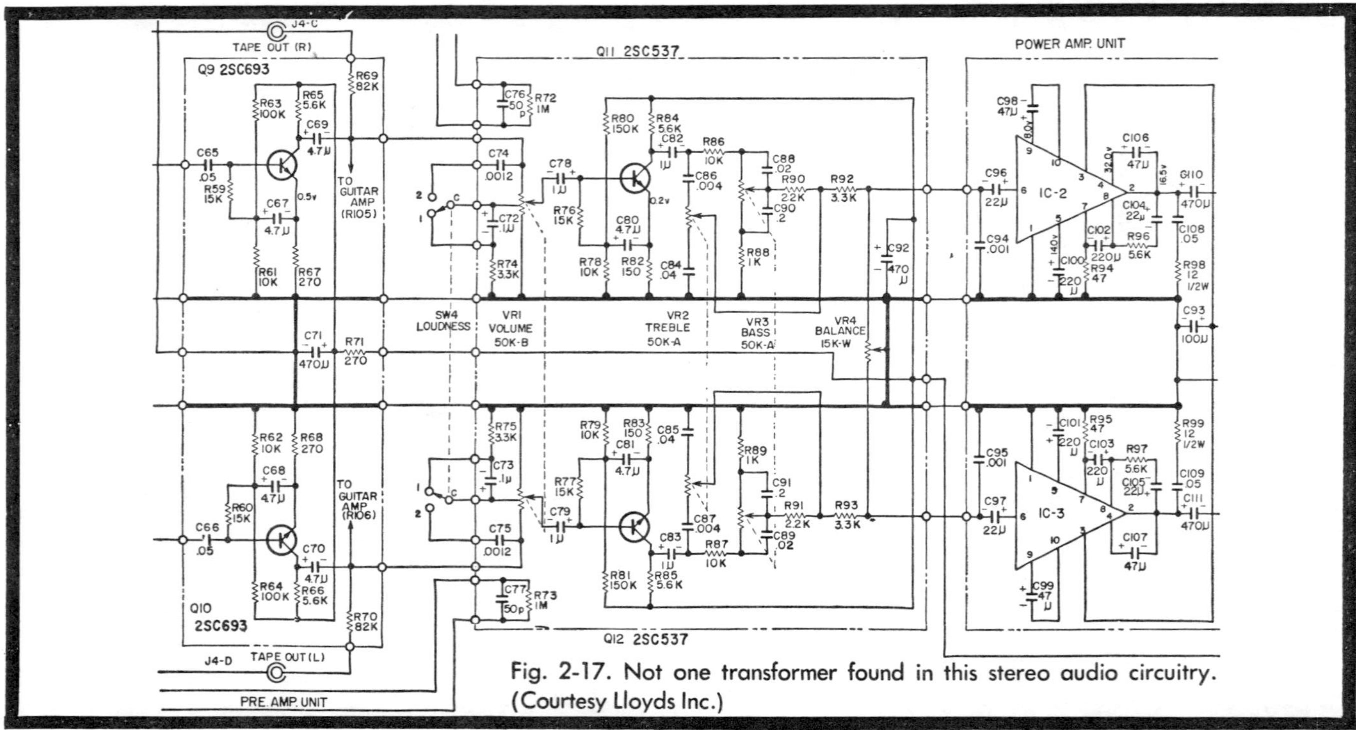

Fig. 2-17. Not one transformer found in this stereo audio circuitry. (Courtesy Lloyds Inc.)

capacitor tied to the speaker system. Signal injection won't work here, but a signal-tracing amplifier can spot the culprit. Also, the electrolytic can be shunted with a good one to see if it is open. Carefully connect a new one across the suspected capacitor with the power off.

Caution: You can easily "pop" transistors by shunting electrolytic capacitors into a circuit. When touched into a circuit, a charged electrolytic may destroy the transistors. One necessary precaution is to always connect a speaker into both channels before attempting to service the audio circuits since a high volume setting with no load can destroy the output transistors.

The physical appearance of a transformerless audio circuit is shown in Fig. 2-18. The absence of transformers contributes to the compactness and lightness of this unit. Most modern 8-track stereo units are of the transformerless amplifier variety.

FOUR-CHANNEL AUDIO CIRCUITS

The four-channel cartridge player's audio circuits are somewhat similar to most stereo players. Of course, there

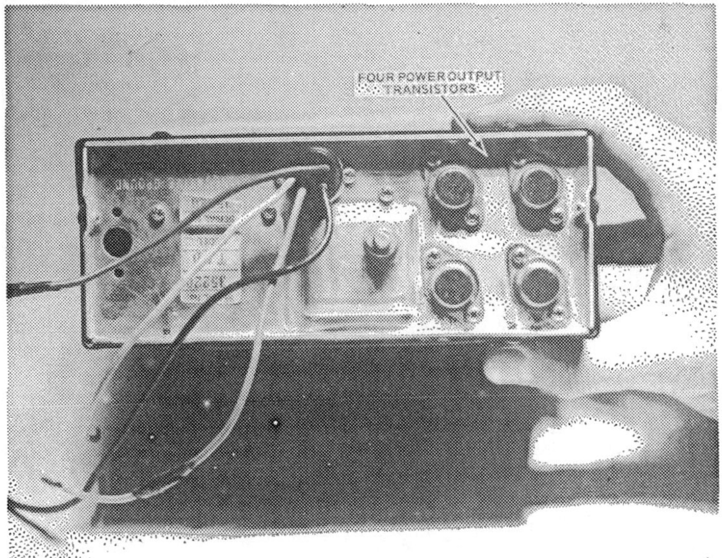

Fig. 2-18. The push-pull output transistors may be found on back side of auto stereo 8 tape player.

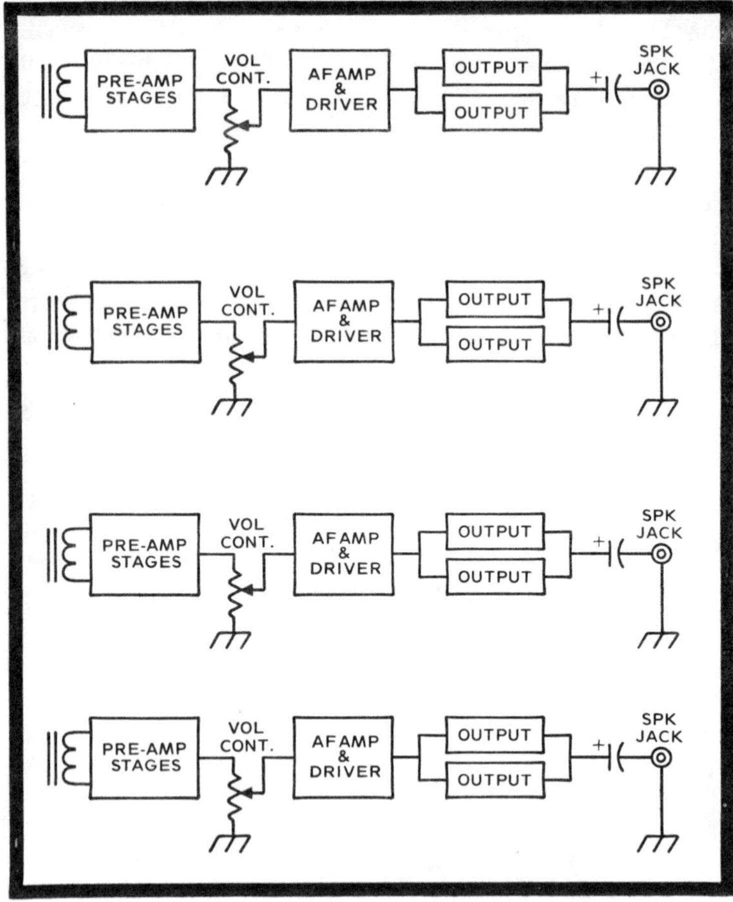

Fig. 2-19. Block diagram of a Q8 stereo player.

are four separate audio channels (Fig. 2-19). There are four separate tape head circuits within the single tape head assembly. The preamp stages may be ICs or direct-coupled transistor stages. When a stereo cartridge is played, only two channels are used. A relay between preamp and audio amplifier stages switches between two and four channels. Generally, push-pull transformerless output stages are used. A typical four-channel auto tape player is shown in Fig. 2-20.

POWER SUPPLY CIRCUITS

You may find half- and full-wave power supply circuits in the inexpensive portable cassette player. Battery-operated

Fig. 2-20. There are a total of 8 power transistors found in this Q8 auto stereo tape player.

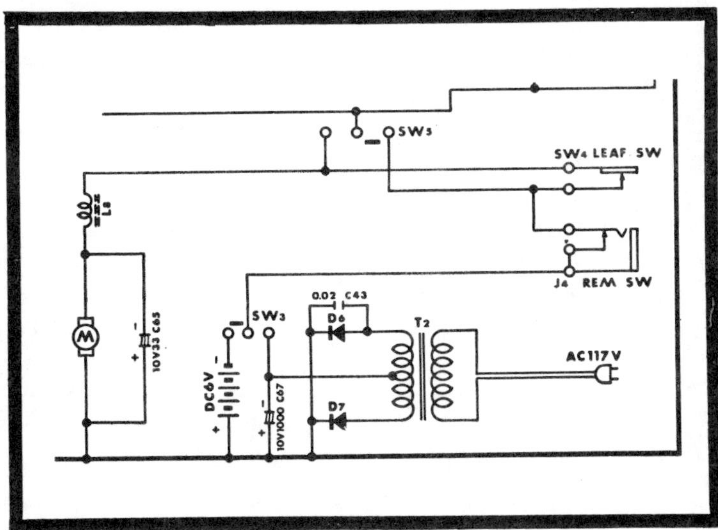

Fig. 2-21. Portable cassette tape recorder power supply circuit. (Courtesy Midland International.)

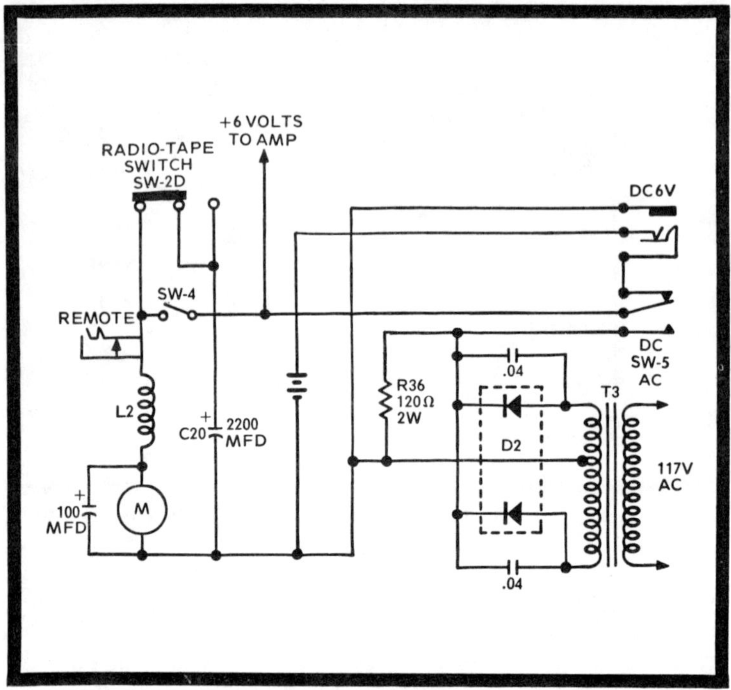

Fig. 2-22. Radio-tape portable cassette power circuit.

units, of course, usually have no power supply at all. In Fig. 2-21 is a full-wave power supply circuit of a portable cassette player. Power transformer T2 steps down the line voltage to rectifiers D6 and D7. The output voltage is filtered by capacitor C67, a 1000 μF electrolytic. SW3 switches the player to either the internal battery or the ac supply. SW3 is a leaf type on-off power switch.

Some cassette units feature "three-way" operation—they can be operated from internal batteries, an external dc source, and conventional power-line voltages (Fig. 2-22). Here we find full-wave rectification with both diodes (D2) in a single component package. R36 is a 120-ohm voltage-dropping resistor. The player is switched to ac or battery operation with SW5. When a battery is inserted into the DC 6V jack, the internal batteries are switched out of the circuit. SW2D is a tape—radio switch and is shown in the tape position. A large-value capacitor filters out any ac ripple of the 6-volt supply.

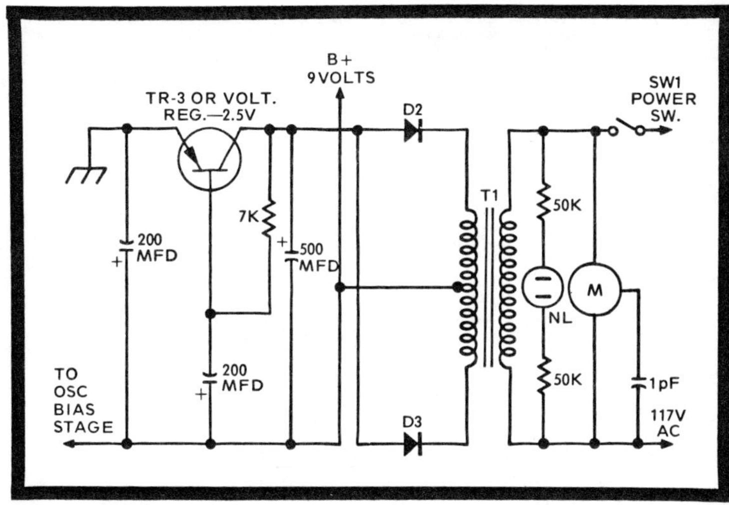

Fig. 2-23. A home cassette tape deck power circuit with transistor regulation.

Most home cassette tape decks employ full-wave rectification (D1 and D2) in the power circuits. A few even have transistor regulators to maintain a constant output voltage (TR3 in Fig. 2-23). Most, however, have conventional full-wave rectification with capacitor-input filter networks (Fig. 2-24). In any case, you'll find cassette power supply circuits are simple, straightforward, and easy to service.

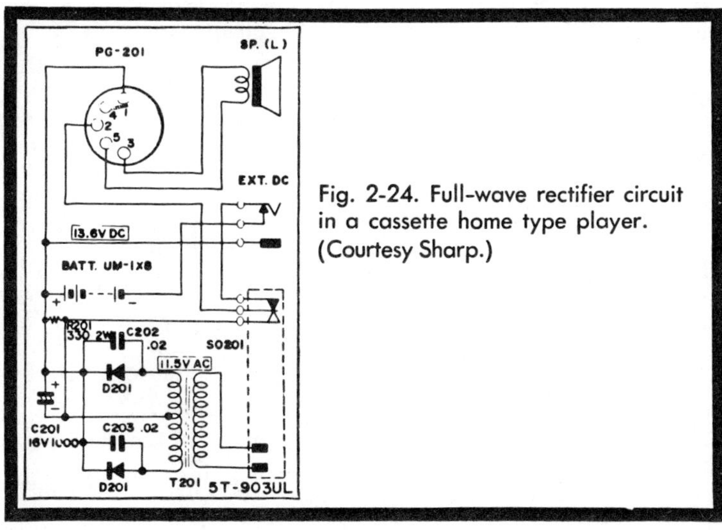

Fig. 2-24. Full-wave rectifier circuit in a cassette home type player. (Courtesy Sharp.)

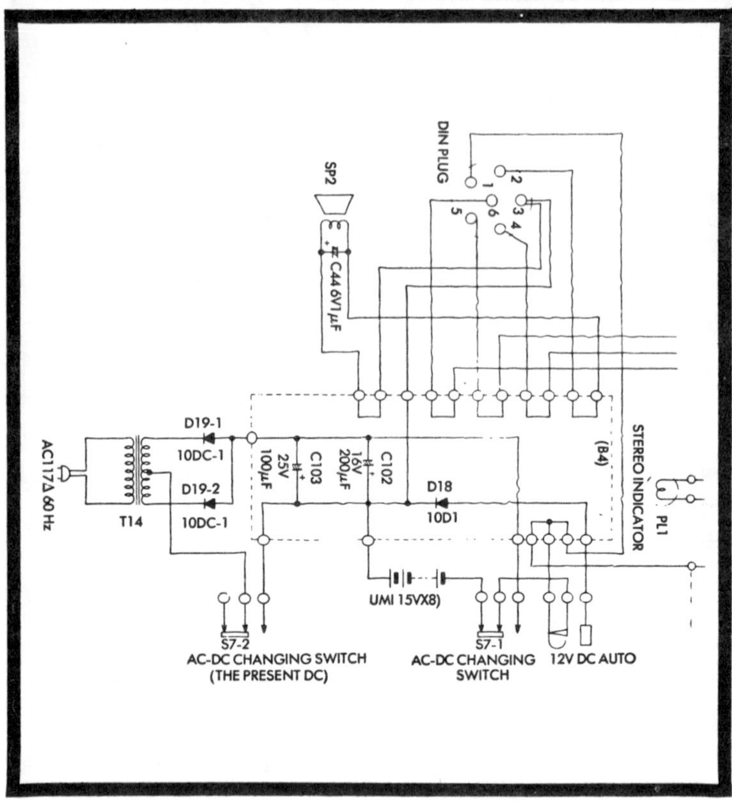

Fig. 2-25. A portable stereo 8 power supply circuit. (Courtesy Automatic Radio.)

In 8-track power supply circuits we find half-wave, full-wave, and bridge rectification configurations. Most portable players will be operated with the internal power supply or batteries. The one shown in Fig. 2-25 is operated from eight D-cells. The ac power supply employs two silicon rectifiers (D201 and D202). A large capacitor (1000 μF, 160V) and resistor form the filtering network. You will find most portable ac power supplies consist of a stepdown transformer, silicon diodes, and a large filter capacitor. In the circuit of Fig. 2-26, the external *auto/boat* jack (J1) supplies power to the player. D2 is a polarity diode that prevents damage to the tape player when voltage polarity is inadvertently reversed.

In Fig. 2-27 another stereo preamp tape deck uses a half-wave rectifier (D1). The stepdown power transformer sup-

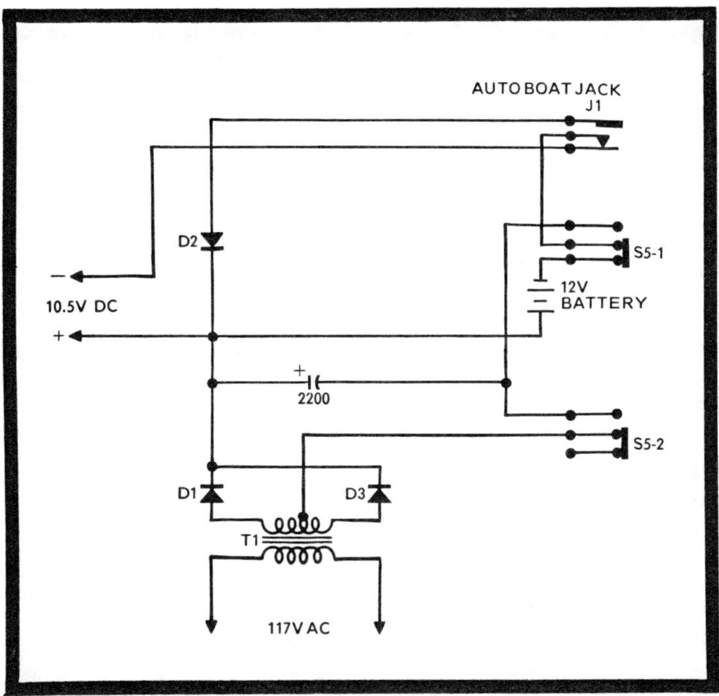

Fig. 2-26. A portable stereo 8 power supply circuit with external boat battery jack (J1).

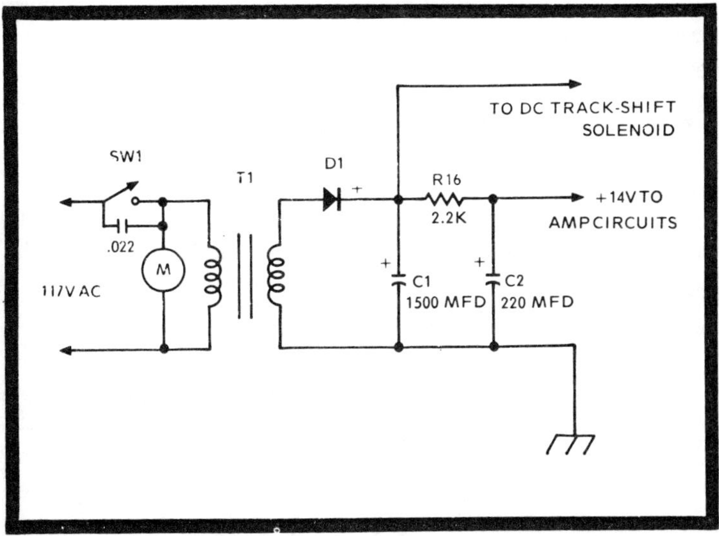

Fig. 2-27. A half-wave rectifier circuit found in stereo 8 preamp deck player.

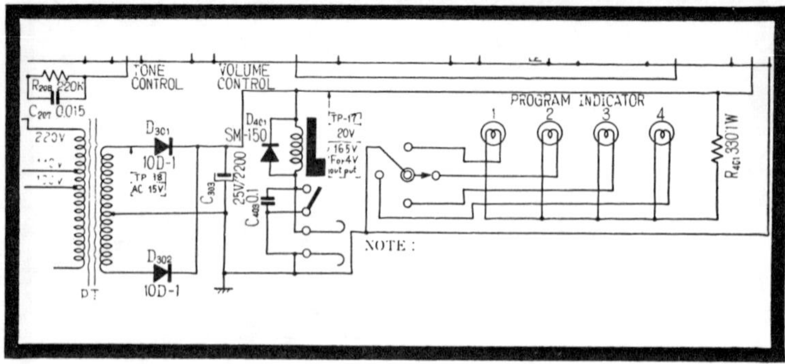

Fig. 2-28. Home tape deck full-wave rectifier circuit. (Courtesy Automatic Radio.)

plies a higher voltage (13.5 ac) to the rectifier. An RC input filtering network furnishes power to the track-shift solenoid and amplifier circuits (18.5V).

With the home stereo tape players, the power circuit may employ both full-wave and bridge rectifiers. You may find a low voltage tap from the stepdown transformer to operate the track-lamp circuits (Fig. 2-28). Bridge rectifier circuits are found in large stereo and combination AM–FM players (Fig. 2-29). Various voltage sources are accomplished with RC filter networks. Sometimes you may find another winding on the power transformer supplying power to the radio dial and tape program lamps (Fig. 2-30).

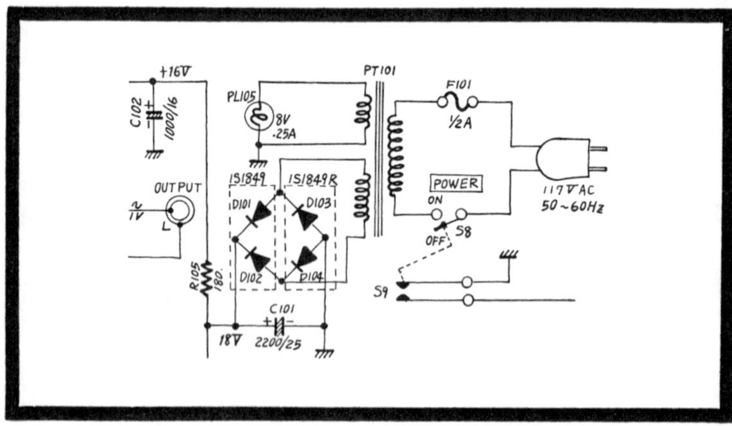

Fig. 2-29. Bridge rectification found in deluxe stereo 8 players. (Lafayette Radio Model RK-890.)

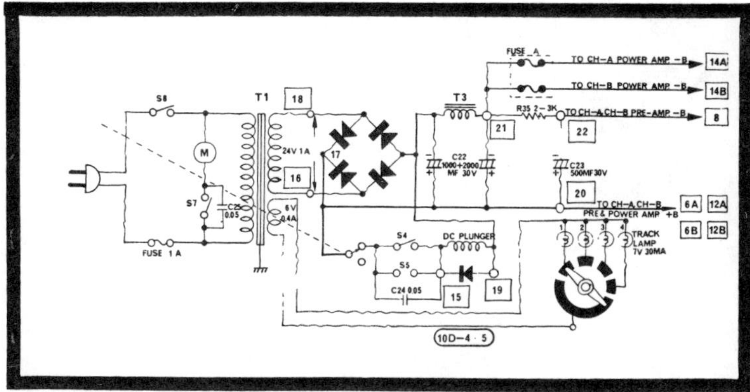

Fig. 2-30. Bridge rectifier circuit with separate dial light winding or power transformer. (Masterwork Model M8601.)

Problems in the tape power supply circuits result in excessive hum, dead set, or weak-audio conditions. Excessive hum may be caused by an open or leaky filter capacitor. A shorted silicon diode may blow the fuse or burn out the primary winding of the power transformer. Since the secondary winding of the power transformer is wound with larger wire, the primary winding usually goes first. It is best to check all diodes in the power supply circuit when one diode is found to be shorted or leaky. Remove one end of the diode to make accurate resistance measurements. A good diode will read a few ohms in one direction and infinite resistance with reversed leads. Most power supply problems can be located with accurate voltage and resistance measurements.

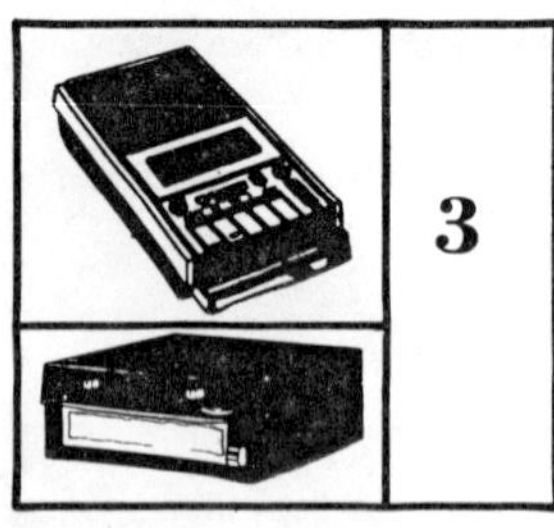

3 Mechanical Section

In this chapter the mechanical sections of cassette and cartridge players are subdivided and described. A lot of the problems found with defective players are mechanical troubles. Even if you are not mechanically inclined, we hope to show how easily most of these problems can be solved.

CASSETTE PLAYERS

Major mechanical components found in the portable cassette player are shown in Figs. 3-1 and 3-2. There are typically many more parts found in the cassette than in the cartridge player. These components are very delicate and miniature. Let's take a look at a few mechanical functions and see how they are serviced.

Pushbutton Switching

Most portable cassette players have a pushbutton function switch to select record, play, rewind, fast forward, and stop positions. In other models you may find a rotary function selector switch, while some models may have a lever-slide switch to do the same functions. The most popular function switch in use today is the pushbutton type.

The most frequent problems found in pushbutton mechanisms are that the buttons will not stay down, release, or switch to corresponding functions. To record, both *play* and *record* buttons must be pushed together. Remember that the *record* button will not lock into position when a recordable cassette is not in place. Check first for a knocked-out cassette spine panel—if the knockout panel is gone, cover the hole with tape. Look for a bent, broken catch or sliding switch lever when a pushbutton will not lock down into position. These switch levers are generally made from metal that

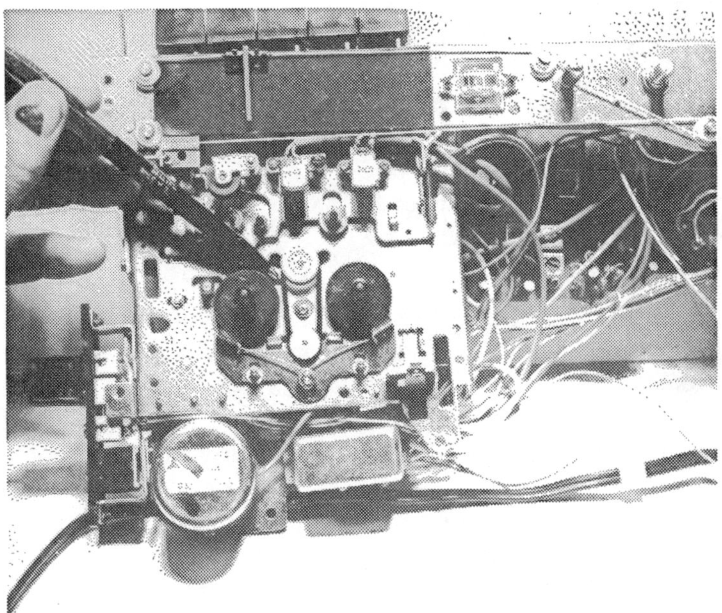

Fig. 3-1. Top view showing major mechanical parts of a cassette player.

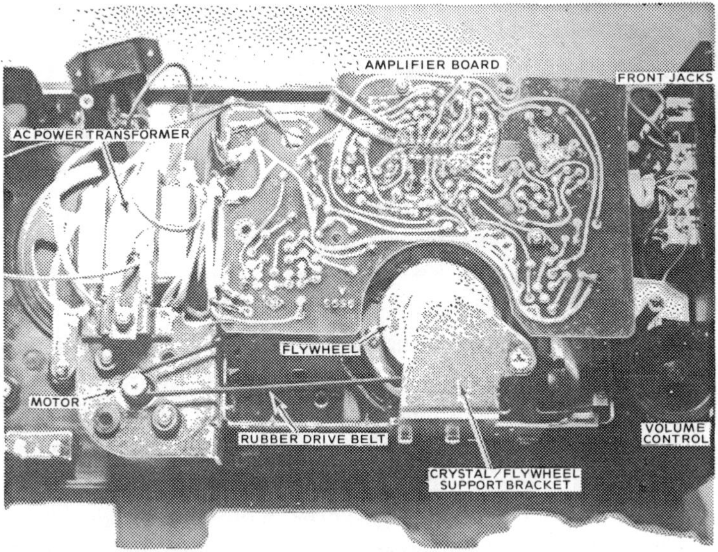

Fig. 3-2. Bottom view showing major mechanical parts of a cassette player.

Fig. 3-3. Pushbutton section of a portable cassette player.

bends very easily. After many hours of use, the lever may bend out of alignment and won't catch properly. Sometimes you can simply bend and form the switching levers back into position. If they are badly worn or bent way out of shape it is best to replace the switch–button assembly. Check the spring tension of the switch lever; poor spring pressure may not let the button lock into position. You can provide smoother switch operation with a thin coat of light grease or oil. Suspect a defective cassette when the stop function is engaged before the cassette has completed its full travel. Sometimes a defective cassette will become sluggish; the drag will then trigger the stop function (or the *reverse* function in the case of automatic reversing cassette players).

Motor Drive

The capstan–flywheel is rotated by a small dc motor and pulley with one or more drive belts. Motors in portable cassette players are powered by batteries or from a rectified power-line (Fig. 3-4) voltage. Some players may have two separate motors—one for slewing (fast forward and rewind) and one for play–record. In some models, the fast forward or rewind operation is accomplished by changing the polarity of voltage to the dc motor. This same operation is accomplished in other models by placing or moving different size pulleys. In some auto cassette players the polarity of the car battery is switched in forward and reverse operations. Fast forward operation in some models is accomplished by applying a higher dc voltage to the motor instead of changing pulleys.

Capstan–Flywheel

The capstan drive shaft is very thin since the cassette's tape speed is only 1⅞ ips. The cassette tape is driven between capstan and pinch roller. The size and weight of the flywheel tends to damp out any variations in motor speed. Since the capstan drive-shaft area is very thin, wow and flutter conditions may result with excessive oxide upon the capstan drive assembly.

The capstan–flywheel assembly is held in position with a flywheel support bracket (Fig. 3-5), which may be anchored at either (or both) end. This bracket contains a lower flywheel bearing, which may be made up of bronze or cast

Fig. 3-4. A small motor assembly found in the cassette tape player.

Fig. 3-5. The cassette capstan—flywheel assembly held into position with metal support bracket. Notice two Phillips side screws and one at right top securing bracket to chassis.

material. You should always remove the capstan–flywheel assembly for cleaning or inspection. First, remove drive belt and bottom support bracket. Very few cassette capstan–flywheel assemblies need replacement. Occasionally, you will locate an excessively worn bearing and the assembly should be replaced. This type of repair may be too costly; check and see if the cassette is worth servicing. If there is a lot of end play between flywheel and support bracket, check for a thrust clearance adjustment screw. (This screw is located in the bottom side of the support bracket.) Adjust the flywheel axis for clearance of 0.1–0.3 mm.

Pinch Roller

The pinch or pressure rubber roller is mounted on a movable support lever (Fig. 3-6). The pinch roller pressure is controlled by a tension spring. This pressure is measured with a spring scale or tension gage and the nominal adjustment will vary between 8 and 16 ounces. If in doubt, check the manufacturer's literature for correct tension spring pressure. If the pinch roller rolls up the tape as a result of part replacement, make the adjustment according to the manufacturer's specifications. A dirty or worn roller may produce wow or flutter conditions. Periodically when cleaning the tape head wipe excess oxides from the pinch roller.

Motor Drive Belt

You may find a flat, round, or square drive belt in cassette players. It's possible to find two drive belts in larger players. The rubber drive belt turns the heavy flywheel and rotates the tape between hub spindles (Fig. 3-7). Speed problems with the drive belt are most frequently caused by oil spots; excessive tape oxide runs a close second. Check to see if the motor belt, pulley, and flywheel are in line. A twisted square belt may produce wow. When a drive belt will not stay on, suspect a stretched belt. Loose drive belts will also cause slow speeds.

Takeup Reel Spindle

The takeup reel spindle takes up the loose tape with the takeup reel and spindle drive pulley (Fig. 3-8). At the very top of the reel spindle is a small plastic cap which slides inside the cassette hub. The plastic cap is spring loaded and

Fig. 3-6. The pinch roller rotates against capstan drive to move tape in play and record positions.

Fig. 3-7. The small square drive belt rotates the capstan—flywheel assembly.

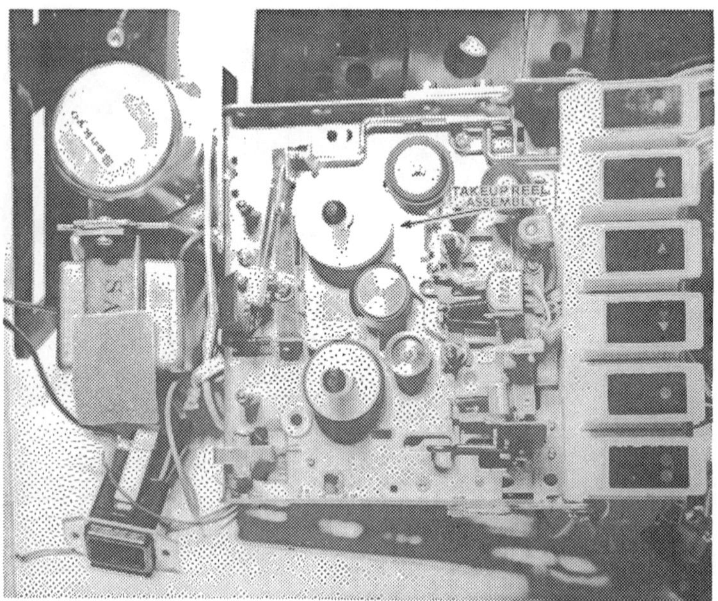

Fig. 3-8. The arrow points out the position of the takeup reel.

compresses when the cassette is loaded. It fits snugly inside of the cassette hub assembly.

The takeup spindle (it may be called a turntable in the manufacturers' literature) may serve two different operations: Besides taking up the tape it may serve in a fast forward operation and be driven by a fast forward idler pulley. Suspect a slow takeup spindle assembly when the tape bunches up in the cassette. Check for oil on the spindle drive surface when the fast forward rotation seems sluggish or erratic. And don't overlook the most common problem of all: a bad cassette.

Supply Spindle

As the tape is being pulled between capstan and pinch roller the supply reel turns uniformly to reel out the tape for this operation. Unless the player has a reversing feature, the supply spindle is always on the left side of the cassette (Fig. 3-9). In rewind operation the supply spindle is driven by the rewind roller and goes in the opposite direction.

The supply and takeup spindles can be removed for cleaning and inspection after first removing the reel caps.

Fig. 3-9. The supply reel is always on the left side of the cassette tape player in nonreversing players.

Fig. 3-10. The cassette brake assembly stops both supply and takeup reels at the same time.

Generally, a slotted *C*-washer secures the cap and spindle assembly into position. Just separate the brake lever from the reel spindles and pull spindles from their shafts. When rewind operation is erratic suspect a dirty, worn, and binding reel spindle. You should also check for a dirty or worn clutch and a binding brake assembly.

Brake Assembly

When the spindles are not rotating, the brake is on. In forward, rewind, and play—record position the brake assembly lifts from both reel spindles. When the stop button is pressed, the brake lever engages the reel spindles and stops the tape (Fig. 3-10). The brake assembly may be a piece of curved metal that fits snugly upon the spindles; but in some units the brake shoe just pushes against the spindles. In stop position, check to see if the brake shoe is pressed against the spindles. When in play, fast forward, and rewind modes the brake shoes on both spindles should be the same distance apart. If not, this adjustment can be made by bending the brake lever assembly. In some units you can adjust the tension spring of braking torque; this should be adjusted in accordance with the manufacturer's specifications.

When the takeup reel does not rotate, check for a defective takeup reel brake. Insufficient disengaging of the brake lever can produce erratic and sluggish speeds. Since the brake shoes ride on the spindles, a dirty or oily brake may contaminate the takeup and supply spindles. Don't forget to clean up those brake assembly areas when cleaning up the tape head. A broken or excessively worn brake shoe should be replaced.

Pop-Up Lid

The pop-up lid is found in expensive protables and some (notably Philips) home tape cassette players. First, load the cassette by pushing it into the lid opening and push the whole assembly down. The lid is locked into position under spring tension. When the pop-up button is pressed the lid pops up with cassette. Most of the pop-up lids are mechanically operated. Problems encountered with the pop-up lid are bent levers, weak springs, and deformed housings. Sometimes the lid will bind, causing erratic operation. You can quickly spot this trouble with telltale rub marks on the sides of the plastic lid.

CARTRIDGE PLAYERS

There are fewer mechanical components found in stereo cartridge players; many are shown in Figs. 3-11 and 3-12. Major problems found in the cartridge player are with the tape head assembly, capstan—flywheel, motor, track-shift assembly, solenoid, and azimuth and height adjustment. Let's tackle each one in order.

Tape Head Assembly

The tape head assembly in a cartridge player consists of a tape head, head support assembly, head holder and support shaft, azimuth and height screws, and tension springs. The tape head may be held in position with a spot weld, or it may be clamped or bolted to the bracket assembly (Fig. 3-13). In too many instances the spot-weld breaks loose, letting the tape head flop around. The tape head is raised and lowered

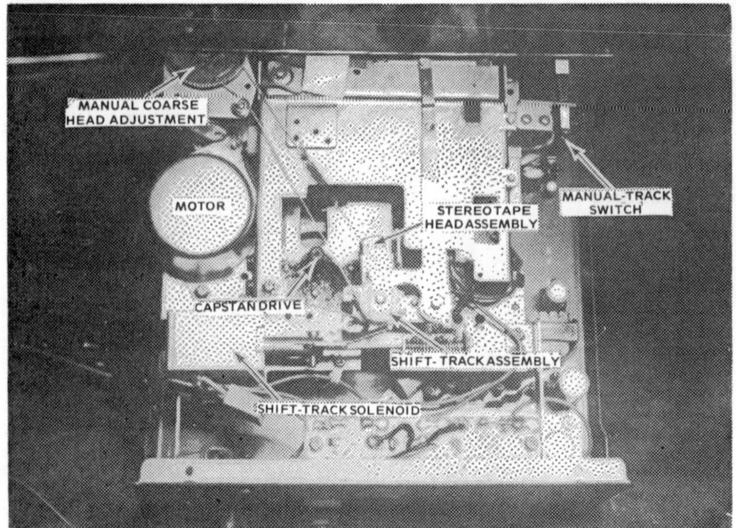

Fig. 3-11. Top view of major components in an 8-track cartridge player.

by a track cam of the ratchet assembly. Some inexpensive portable models may have a mechanical knob you can turn to raise or lower the tape head. When crosstalk is encountered the head is probably out of adjustment. Tape head adjustments are quite critical and a complete rundown is given in

Fig. 3-12. Bottom view of major components in a stereo 8 player.

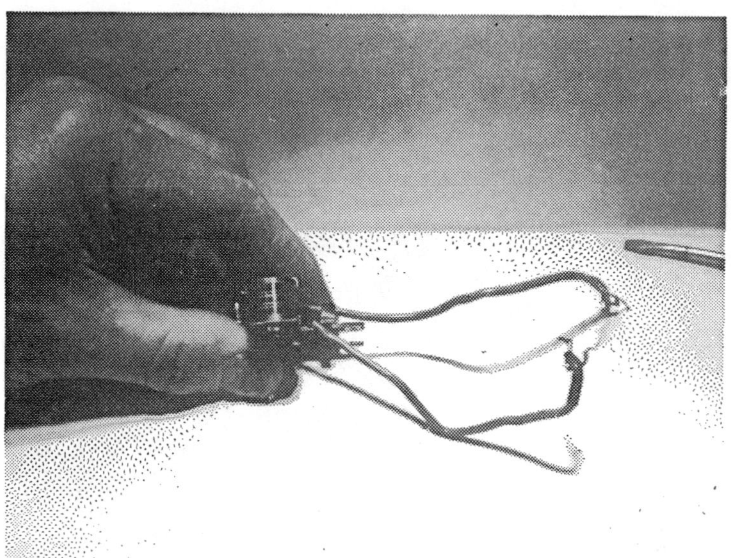

Fig. 3-13. Closeup of tape head and mounting bracket. Notice the two separate shielded cables coming from tape head.

Chapter 4. In some auto players, the tape head is adjusted by a fine tuning adjustment on the front panel. In this case, the tape head assembly is adjusted with a dial cord connected to a pulley and knob.

Problems caused by the tape head assembly are related to height adjustment and track changing. Improper head adjustment can produce crosstalk and distortion. When the tape head will not shift properly, check for a defective track cam located upon the ratchet assembly. It's possible the tape head may be too tight, thus preventing the head from going to the lower position. Likewise, loose adjustment of the tape head may not let the head cover all four tracks. A broken or missing tape head pin may not let the head move at all.

Capstan Drive

The capstan—flywheel is driven by a motor belt, and the drive shaft rotates between tape and rubber roller, thus pulling the tape through in a continuous operation. The motor speed and the circumference of the capstan shaft determine the tape speed. Generally, the spindle or shaft is constructed of hard steel and the flywheel from cast metal (Fig. 3-14).

Fig. 3-14. This capstan drive shaft is from an early Lear-Jet model. Actually, the motor field turns in this motor. The whole assembly turns with the bearings at the top and bottom of the capstan drive.

Almost all of the problems found with the capstan–flywheel assembly are related to speed.

A dry or frozen capstan bearing may produce erratic or slow speeds. The capstan bearings may be located at the top, middle, and end areas. Check the capstan bearings for worn or dry areas. When excessively worn bearings are found, replace the entire capstan assembly (but first check the cost of these assemblies to determine whether or not the unit is worth repairing). A scored or rough capstan drive will pull tape from the cartridge. Sometimes, you may solve most speed problems with a good cleanup and proper lubrication.

The capstan–flywheel may be dismounted by removing the metal and bearing plate. Many flywheels are held into position with a *C*-washer at the top, near the capstan drive area. If the *C*-washer is located near or under the bottom bearing it may be rough to remove. Use a thin tool with a bent tip to pull the *C*-washer out. Remove the steel pin running through a slot of the capstan bearing assembly to release other flywheels. You may have to remove several components before the capstan–flywheel is free.

A dry nylon washer found at the very top of the capstan drive bearing may produce *wow*. Sometimes this washer is located under a plastic cap. Other wow and speed problems can be caused by oil on the flywheel and capstan drive surfaces. If the capstan–flywheel has a shiny surface the drive belt may be slipping. Check also for small rubber particles stuck to the flywheel surface.

If you hear a scraping noise when the flywheel is rotating, the flywheel has probably dropped down against the support bracket. More than likely, the flywheel has worked its way down the steel drive shaft. Don't be too surprised to find these conditions in new players; it's happened many times. If the flywheel takes a few weeks to work down the spindle, sometimes you can repair it. First, line up the flywheel upon the drive shaft and place a layer of epoxy under the flywheel area. Let the cement set up over night before installing the flywheel assembly. In most cases the capstan–flywheel asemblies are quite durable and only the capstan bearing assembly needs replacement.

Motor Assembly

Naturally, all motors found in the auto stereo players are dc operated with armature and brushes. In the portable players the dc motor may be operated from batteries or the power line. Virtually all motors found in home players are ac induction motors. The smaller induction motor (Fig. 3-15) causes very little trouble compared to the dc motor. In operation, the motor provides tape drive indirectly by virtue of the belt coupling to the flywheel.

A defective motor may produce slow, intermittent speed or it may "freeze up" and not rotate at all. When the motor freezes, check first for dry bushings or bearings. You can quickly check a motor's condition by turning the player on and rotating the motor shaft between your fingers. If there is no torque or movement, check for correct voltage at the motor wire terminals. In case the motor starts when the pulley is rotated suspect a flat armature. A dead motor may have brushes not making contact or open field windings.

Erratic or intermittent motor speed may be caused by a bad scratch located inside the motor. Some dc motors have

Fig. 3-15. The small motor drives the capstan—flywheel with a rubber belt within a stereo home player.

internal centrifugal speed governors, of which the contacts become dirty. Sometimes you can get by nicely if you clean these points with postcard paper and cleaning fluid. Generally, these repairs should only be made if a new motor is not obtainable or the repair does not warrant a new motor. You should replace the motor assembly if the points or switch will not restore proper speed.

Excessive playback speed may be caused by a stuck regulator switch inside the motor or a defective electronic speed circuit. Simply tap the motor case with a screwdriver and notice if the motor speed decreases. Sometimes when pushing the manual selector switch, the motor will speed up, indicating a defective switch contact. If this is the case, grip the fast-moving motor pulley and slow the motor down. The motor is definitely defective if the motor resumes normal speed under this condition; it should be replaced.

When a motor is suspected of throwing drive belts, check for a loose or deformed motor pulley. See if the motor pulley is in line with the capstan—flywheel. A loose or stretched belt will tend to fly off the flywheel. You should check the motor mount for loose or unstable operation. A motor mounted inside a rubber or metal container can be repaired by wrapping several layers of tape around the motor. Then tighten the motor assembly and see if the motor mounting is secure.

A noisy motor may have excessively worn top bearings. The top bearing will usually wear out first because of tension from the drive belt. Remove the drive belt and check for end play of the motor shaft. Sometimes a drop of oil may cure the noisy bearing; but generally this repair is only temporary. Replace the motor when excessively worn bearings are noted.

Track-Shift

The track-shift or channel change assembly shifts the tape head to a different set of tape tracks. The track-shift assembly consists of a solenoid, panel, changing cam, shaft, program indicator arm, lever, and springs (Fig. 3-16). Track indication is shown by a mechanically driven dial or four channel lights. A breakdown of any of these components may result in no channel change, erratic operation, and improper channel indication. The track-shift assembly changes chan-

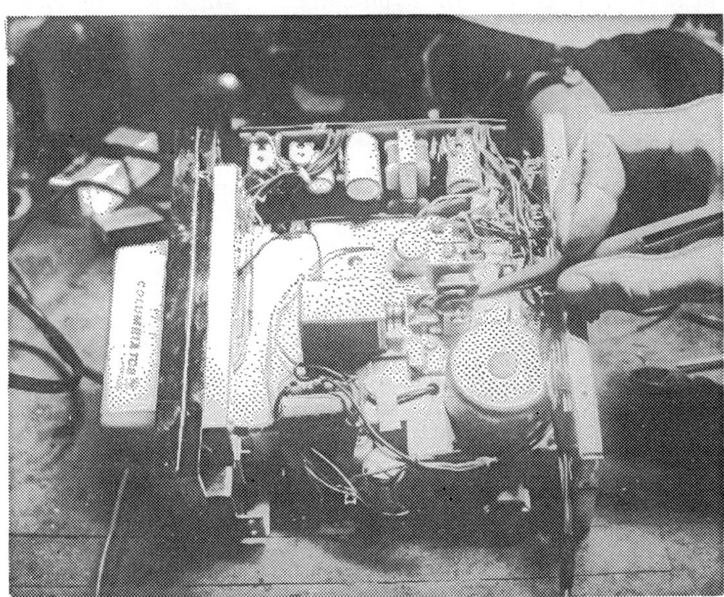

Fig. 3-16. A cartridge player track-shift assembly.

Fig. 3-17. The arrow points to the automatic program change switch. Generally you will find one side of the switch grounded and wired in parallel with the manual program switch.

nels when the manual channel button is pushed or when the program is finished. The automatic track-shift function occurs when a length of metal foil (on the tape) makes contact between two switch blades.

The automatic switch consists of two stationary metal pieces and rubs against the tape at all times. When the tape contact is made the solenoid is energized, pulling a lever panel that rotates the head cam. The head cam may raise or lower the tape head assembly. Simultaneously, the track dial assembly is rotated, indicating which program (set of tracks) is playing.

A complete cleanup and lubrication of the ratchet and solenoid assembly will solve most sluggish track-shift problems. Sometimes excessive oxides on the automatic change switch may prevent the solenoid plunger from seating properly; this results in jamming of the shift mechanism. Improper seating of the lever panel and head cam may produce improper tripping. Erratic channel operation may also be caused by a dirty, oxide-caked program switch. Check the manual program switch for dirty or poor contacts. If the track-shift assembly functions manually and does not operate with the automatic program switch, suspect a broken or worn automatic switch blade (Fig. 3-17). When the player is turned on and you find the solenoid energized, suspect a grounded or shorted automatic program switch.

Solenoid

The solenoid assembly operates the lever panel and head cam. When the manual or automatic program switch is on, it supplies a voltage to the solenoid circuits. A metal plunger pulls inside the solenoid to move the lever panel. You will find most track-shift problems are caused by the solenoid and its surrounding components.

When the automatic or manual switch is engaged, a shorted suppressor diode across the solenoid winding may blow the main fuse (Fig. 3-18). Actually, the shorted diode places the positive battery terminal of the car to chassis ground. Sometimes the plunger will not seat properly inside the solenoid, resulting in improper track changing. Check the

Fig. 3-18. The pen points to a suppressor diode wired across the solenoid.

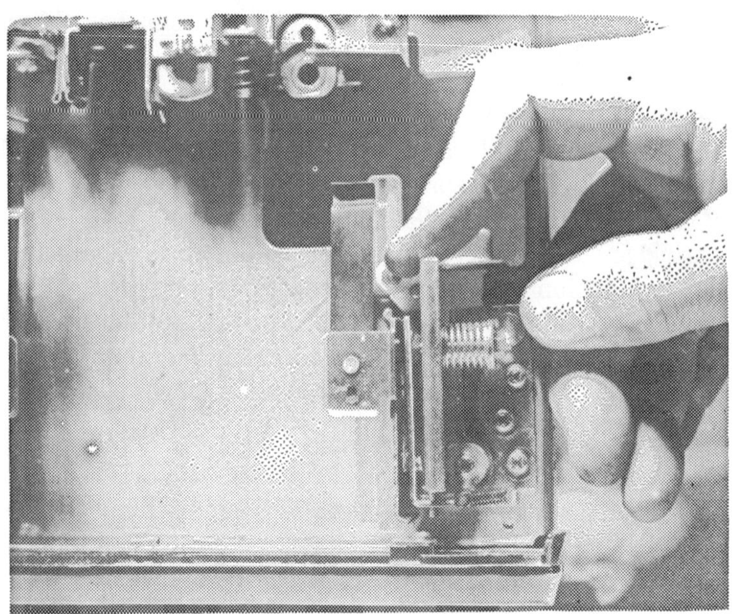

Fig. 3-19. Here is a mechanical cartridge ejector assembly of a cartridge tape player.

plunger, lever, and spring for complete operation. A warped or worn plastic sleeve inside the solenoid may not let the plunger seat properly. You can repair the sleeve by enlarging or drilling the hole the same size as the plunger. If the solenoid is obtainable, it is best to replace it.

Suspect a defective solenoid when it appears scorched or burned. In some cases you may find the solenoid burned to a crisp. Generally, this condition is caused by a jammed track-shift assembly or shorted automatic program switch. You will find the resistance of the solenoid quite low and made up of many turns of large-diameter wire. Very few solenoids open, but they do operate quite warm.

Check the tension of the pullout spring when the plunger will not return to normal position. In some auto players the solenoid has a plunger adjustment for correct operation. A bent arm lever and dry cam may not let the head shift to another track. Improper track-shift may also be caused by a worn panel or cam, or a bent arm lever and panel spring. Foreign material such as pieces of broken tape, gum wrappers, and car keys serve well to prevent the player from changing channels automatically. (It is surprising how these items are collected in the cartridge opening!)

Cartridge Ejector

The mechanical cartridge ejector consists of nothing more than levers and springs. When the cartridge is inserted it is locked into position under spring tension (Fig. 3-19). The cartridge can be ejected by pushing a button to relieve the spring tension and kick the cartridge outward. Most mechanical ejection problems are caused by bent levers and weak springs. Check the ejector assembly for worn or dry lever areas. You will find the lever material bends easily and sometimes can be bent back into position. If not, replace the entire ejector assembly.

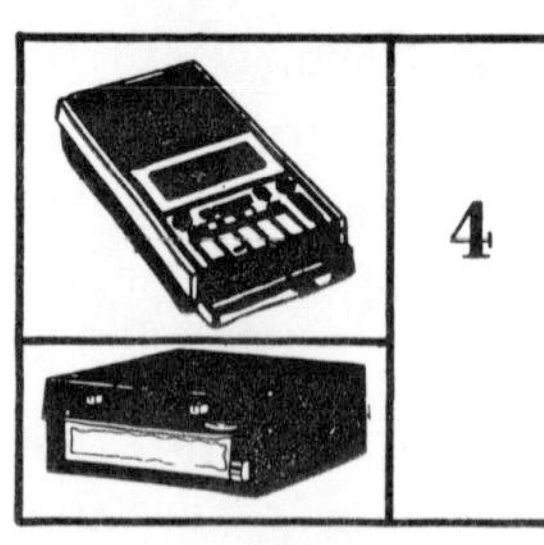

4 Testing and Adjustments

It is not difficult to test and adjust the various electronic and mechanical components of cassette and cartridge players. Prerecorded test tapes are used to adjust tape head and to check operation of amplifier sections. You can keep the tape player in tiptop shape by making minor adjustments. The first section of this chapter deals with test tapes and adjustments of the cassette player. Cartridge players are dealt with in the latter part of the chapter. The adjustments are quite simple, so let's get at them.

CASSETTE SYSTEM TESTING

There are many cassette test tapes on the market to check out the various functions of cassette players. You should select one that incorporates head alignment, frequency response, sweep frequency, and intermodulation tests. The photo of Fig. 4-1 shows the GC *Audiotex* test tape that we'll be referring to in the test that follows.

Head Alignment Tilt

This test (7500 Hz) is for proper angular positioning of the playback head with relation to the tape. Most recorders have screws located near the playback head for this adjustment. Adjust screw until the meter shows a maximum deflection or until the bulb glows at maximum brightness.

Frequency Response and Equalization Test

The following series of tones (70—8000 Hz) are for checking the overall frequency response of the tape playback system. The first tone, 1000 Hz, is used to set playback level. Note meter reading or adjust volume until the bulb barely glows, but do not touch the setting after that. With proper equalization the meter indication (or brightness of bulbs) should be approximately the same at all frequencies. Output variations may be compensated for in part by slightly resetting the bass and treble controls. On lower priced players, the playback level may drop at the extreme lower and upper ends of the response range. Indeed, sometimes the 40 Hz note cannot be heard at all.

Sweep Frequency Test

The next test is a continuous frequency sweep down in frequency from 8000 to 70 Hz. It will clearly indicate any serious peaks or dips your system may have. This sweep frequency test will also indicate any serious resonant peaks in the speaker or listening room which may detract from listening pleasure.

Intermodulation Distortion Test

Two frequencies, 7500 and 100 Hz, are mixed in a 4:1 ratio at 75% modulation at – dB. Only two frequencies should be heard. Any buzz or undesirable "products" may indicate trouble—poor circuitry, dirty heads, bad speakers, etc. If

Fig. 4-1. Using a cassette test tape for making head alignment and frequency response test in cassette tape players.

Fig. 4-2. Hartak's X87 Torqette is a specially made cassette for use in determining a player's instant torque.

available, the use of intermodulation distortion test equipment is recommended for accurate measurements.

Channel Identification and Separation

This test is to identify each channel and to help in placement of speakers.

Cassette Instant Torque Test

The X87 Torqette is a torque-test cassette with a built-in gage that allows the technician to measure the amount of takeup reel torque. This adjustment is very important as misadjustment can cause chronic tape-spillage troubles. The Torqette is inserted into the machine, which is set for the normal play function. The torque is then read directly from the cassette gage (Fig. 4-2).

Insert the Torqette with the dial side up and press play button. Torque below 30 is too low and may cause the takeup hub to stall, the source of such problems as tape snarling, cassette jamming, and tape-fouled capstan. High torque (above 60) will load the motor and may cause wow and other conditions (broken, stretched, or otherwise deformed tapes).

You may also find the batteries in portable tape players have a shorter life when torque is excessive. The recommended torque is between 35 and 60 grams; and the ideal is 40–45 grams.

An erratic and wide-swinging gage indication may be due to slipping tires or poor clutch surfaces. A clutch may change after a tape or two has played. Clutch surfaces must be clean and free of any foreign contamination such as oil. It's best to take a peek at the clutch surface to see if it appears worn or oily. A defective clutch should be replaced before making further tests.

After all drive surfaces are cleaned make certain that the slipping action is confined to the clutch assembly. On rare occasions, some mechanisms may give slightly higher or erratic readings following cleanup and adjustment. In such cases it is recommended that the *Torqette* be inserted and the machine set to the play position for a few minutes without power applied. Switch the player on and off and take several readings.

It is not feasible to repair many clutches because of the methods used in construction; replacing it may be the best bet. Some are held together with a small clip (which itself can be serviced to a limited degree). If the torque is low, the spring may be slightly stretched—assuming that the shaft and the felt pad are in good condition.

Be careful when performing this operation, since an overstretched spring would be ruined and cutting is *not* a recommended procedure. Check the torque in a new cassette player and notice the clutch action; even new ones can have poor torque action. A good clutch will not damage a good cassette.

The *Torqette* can be purchased from Hartak Instrument Co., Box 176, Aurora, Colorado 80010.

Azimuth and Height Adjustments

The head height adjustment is fixed in most cassette players. The head azimuth adjustment point is located on one side of tape head (Fig. 4-3). Generally, a counterclockwise turn of the azimuth adjusting screw will tilt the head upward, and a clockwise adjustment will lower the tape head tilt. This

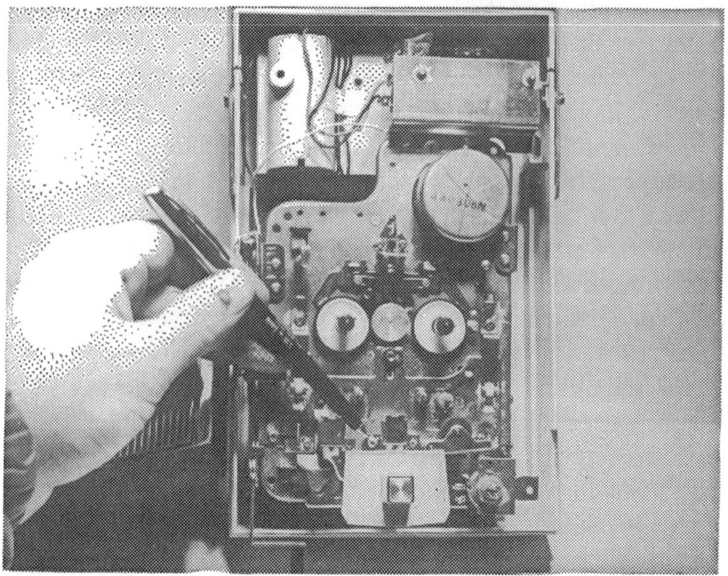

Fig. 4-3. Location of azimuth adjustment in a cassette tape player.

adjustment should be made with a prerecorded test cassette in place; the azimuth screw is to be adjusted for maximum output. Follow the same procedure as outlined for stereo 8 tape head adjustments in this same chapter.

Testing Record–Play Head

The record–playback head is switched in and out of the circuit with the record–play switch. You may find separate record and play heads in a few of the larger cassette players, but the portables, automobile types, and most decks use one head for both record and playback functions.

A defective head coil winding may increase in resistance or appear open. Check head continuity with ohmmeter. This resistance may vary from 150 to 600 ohms. Check the manufacturer's literature for correct resistance. When in doubt substitute another head. On stereo machines, checking a suspect head is simple; it's merely a matter of reversing the two shielded leads connected to the stereo record–play head.

With a distorted channel or no audio in record position, suspect a defective tape head, or trouble in the switching or bias circuits. A dirty, oxide-packed tape head will give the same results.

Check the recording bias across the tape-head terminals with an audio VTVM (record position) as shown in Fig. 4-4. Low bias voltage (or no bias at all) will indicate a defective oscillator section. An intermittent recording can also be traced to a defective bias oscillator circuit. Record bias adjustment should be made according to the player manufacturer's specifications.

Checking Erase Head

When the erase head will not completely erase a previous recording, excessive noise and distortion results. With no erase at all, suspect a shorted or open erase head. A dirty or oxide-caked erase head will cause the same symptoms. Be sure to check for a dirty head before attempting to remove the erase head. Also, check the position of the erase head; if

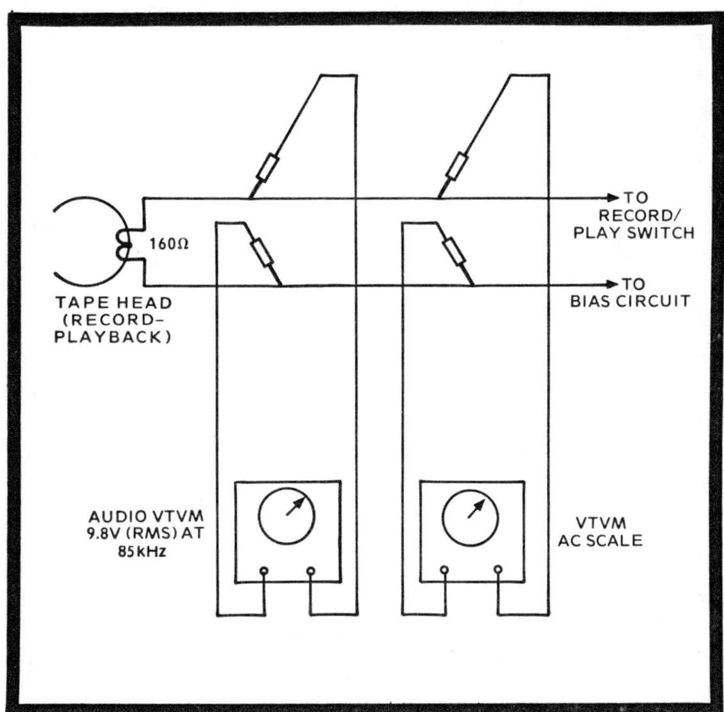

Fig. 4-4. Two methods for checking bias recording voltage at tape head.

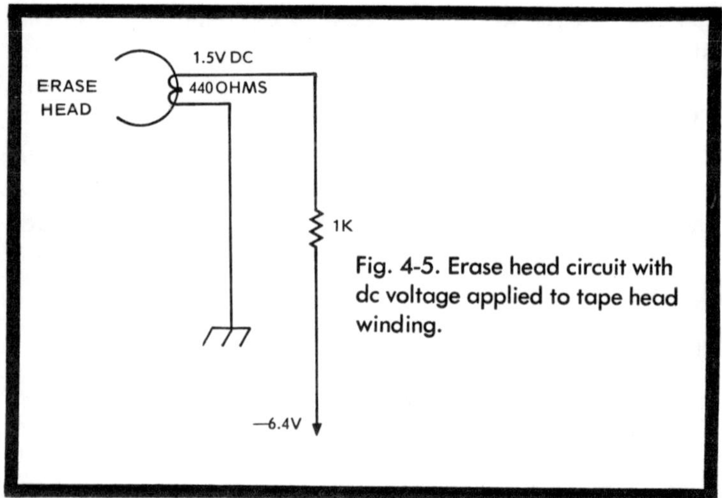

Fig. 4-5. Erase head circuit with dc voltage applied to tape head winding.

it is not in the correct line of the tape path, all you'll have to do is shift it slightly.

The erase head's resistance may vary from 1 to 500 ohms. In small reel-type tape recorders a piece of magnet was swung against the tape to erase or clean previous recordings. Most cassette erase heads are excited with erase current or voltage. A dc voltage in some circuits is switched from the power supply source and applied across the erase head. Generally, the erase head resistance is quite large (above 400 ohms) with direct voltage applied (Fig. 4-5).

When the erase head current is fed from a bias oscillator circuit, the voltage can be measured with an audio VTVM or high-impedance ac voltmeter across the erase head terminals. The erase head resistance in these circuits may vary from 1 to 10 ohms. The erase head current can be measured in this type of circuit by inserting a microammeter in series with the ground lead of the erase head winding. Also, some manufacturers recommend inserting a small resistor in series with the erase head winding and measuring the voltage drop across the resistor. Current can be calculated in this manner by applying Ohm's law: $I = E/R$.

Tension Adjustments

The following cassette pressure adjustments should be made according to manufacturer's literature:

- Pinch-roller pressure
- Fast-forward and rewind torque
- Brake assembly adjustment
- Driving-pin pressure
- Guide screw adjustment
- Adjustment of intermediate idler wheel.
- Takeup torque.
- Cassette insertion force.

CARTRIDGE SYSTEM TESTS

There are many different cartridge test tapes on the market to make head alignment. Select several test tapes for azimuth, height, crosstalk, and auto track switching tests (Fig. 4-6). Misalignment can cause undesirable crosstalk (music coming from wrong track) frequency loss, and distortion. Listed below are several RCA tapes recommended by tape player manufacturers and used by many technicians in the field:

RCA 321 - Azimuth and head height alignment.
RCA 324 - Frequency response.
RCA 326 - Head elevation.

Fig. 4-6. Checking cartridge players with a test tape.

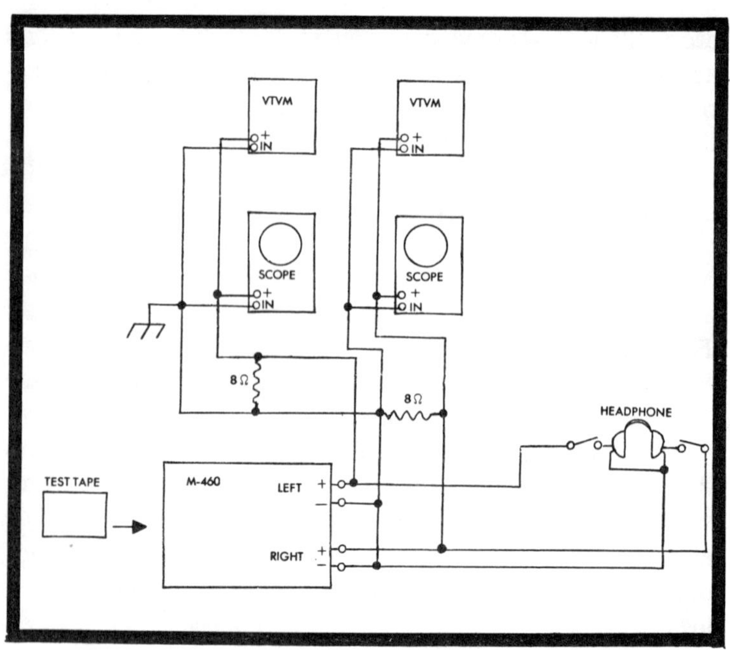

Fig. 4-7. Checking head alignment and frequency response with test tape.

RCA 327 - Crosstalk and track adjustment (even-numbered tracks)
RCA 328 - Crosstalk and track adjustment (odd tracks); separation
RCA 343 - Azimuth adjustment
RCA 356 - Auto track switching test

Test Tape Hookup

A VTVM and scope may be connected to the speaker output terminals for accurate adjustment indication (Fig. 4-7). Connect an 8-ohm, 10-watt resistor across the speaker terminals as a load resistor. Although two sets of test instruments are ideal, as shown, only one set is needed for accurate adjustments. In fact, a VU or balance meter may be used instead of the VTVM or scope as signal indicator.

Head Adjustment Points

After the tape head has been properly cleaned use a test tape and check the head alignment (Fig. 4-8). Remember crosstalk and excessive distortion may result from poor head adjustments. Crosstalk is caused by simultaneous reproduction of two audio sources; another channel is "busting in" on the selected channel. Remember, excessive noise or distortion may be caused by a magnetized tape head. Try another cartridge that is known to be okay.

Head Height. Height adjustment raises or lowers the head vertically so that the head pole pieces are positioned exactly in line with the prerecorded tracks on tape (Fig. 4-9). This is accomplished by adjusting the head elevation screw in or out as required. Use a suitable test tape (RCA 321 or 327) when making this adjustment.

Head Height Screw Location. The head height or elevation screw is generally located on top of the tape head assembly (Fig. 4-10) and near the front. Some head height screws are located directly underneath, raising or lowering the tape head; while in a few models the height screw is located on one side (Fig. 4-11A). Make sure you have the correct adjustment screw before attempting head alignment. As Fig. 4-11B shows, there isn't much discernible difference

Fig. 4-8. Location of height and azimuth screw adjustments in an auto cartridge tape player.

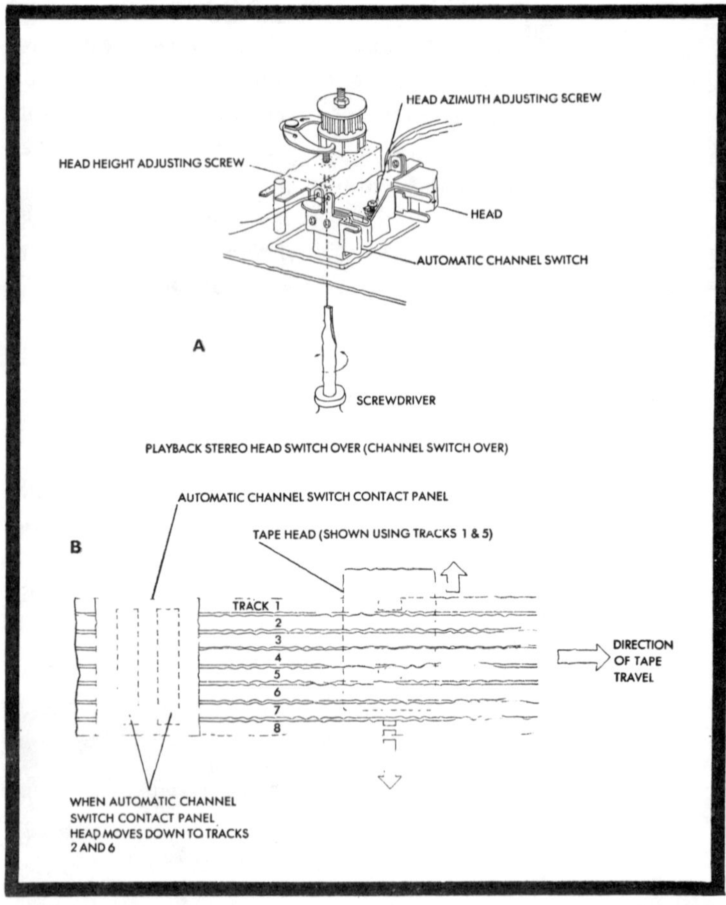

Fig. 4-9. Sketch A shows typical head height alignment point; the screwdriver is pictured for clarity. The tape track layout, track-switch configuration, and head alignment are shown in sketch B. (Courtesy Allied/Radio Shack.)

between the height and the azimuth adjustment screws. In some players the tape head adjustment screws can be located by pulling out plastic plugs or removing a small metal cover without having to remove the metal housing from the tape player. Remember, the head height screw is to eliminate crosstalk or cochannel interference. One last caution here: Be careful not to mistake the head height screw for the head tension or channel indicator screw located on top of some tape head assemblies.

Head Azimuth. Azimuth adjustment provides a means for rocking the head angularly to obtain maximum frequency response; this adjustment is also noted as head tilt. It is done by alternately loosening and tightening the right and left azimuth adjustment screws. For example, to rock the head to the left, slightly loosen the azimuth adjustment screw. To rock the head to the right, slightly tighten the azimuth adjustment screw. After each azimuth adjustment be sure to recheck head height. Repeat the height and azimuth adjustments for optimum head location. Use a suitable test tape (RCA 343) when making this adjustment.

Azimuth Screw Location. The azimuth screw is always located to the rear and one side of the tape head. It may be located either on the right or left side. Here are several ways you may want to adjust the tape head:

Head Adjustment Equipment

You can use a VTVM or VOM to perform head alignment. You may want to use the scope or a level meter connected across the terminated speaker leads as a signal indicator.

Fig. 4-10. The pencil points to a typical cartridge player's height screw adjustment.

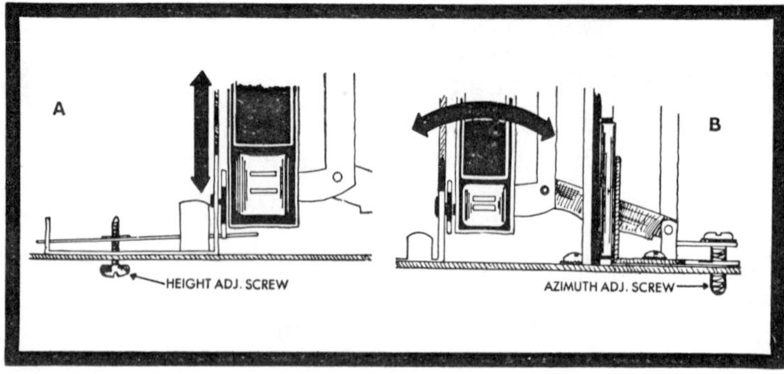

Fig. 4-11. In some stereo 8 tape players the height adjustment screw is located to the side of tape head assembly. (Courtesy Automatic Radio.)

The speaker leads should be terminated with a load representative of the loudspeaker itself. An 8-ohm resistor with a 5- or 10-watt power rating will serve nicely. (You may use the speaker as a load, of course; but you'll have to contend with the sounds of the program—which may well be disconcerting.) You need a bench dc supply to power automotive-type players, of course. This leaves only the head alignment test tapes and a small screwdriver for making the adjustments.

Head Adjustment Sequences

When the head is in the highest position, the pole piece and the tape guide should be parallel at the top edge. Most tape head height adjustments are made with the selector switch set for program 2 (head gaps on tracks 2 and 6). Adjust the height screw for maximum output. Set azimuth for maximum output also; then go back and repeat the height adjustment. This back-and-forth tweaking allows adjustment-point interaction to be minimized. Check for uniform output in all four program channels using the 1 kHz portion of test tape. Then check adjustment on all channels for perfect alignment and no crosstalk interference.

A typical azimuth and height adjustment (using test cartridge RCA 321) is given in the following steps.

1. Connect the VTVM or multimeter to 5V or 10V (ac) range and load resistor to individual right and left output circuits of the stereo tape player (Fig. 4-12).

2. Select program 2 (tracks 2, 6). Set the bass and treble controls to midrange.
3. Turn the height adjusting screw for minimum deflection of left-channel meter.
4. After height adjustment has been completed, turn the azimuth adjusting screw for maximum right-channel meter deflection.
5. Repeat height and azimuth adjustments alternately, since height may change by turning the azimuth adjustment.
6. After all adjustments have been completed, securely lock the height and azimuth screws. (A dab of paint or glue will serve to hold the screw in place.)

Note: A 1000 Hz signal is recorded on each side above and below track 2 (left channel) of the RCA test cartridge. Track 6 (right channel), with an 8000 Hz signal, is used for azimuth adjustment.

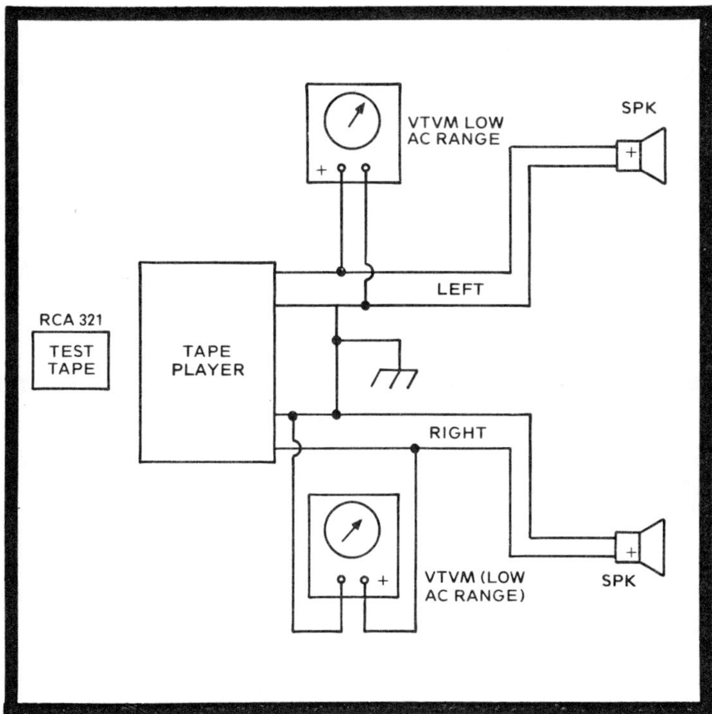

Fig. 4-12. VTVM hookup when using RCA 321 test tape.

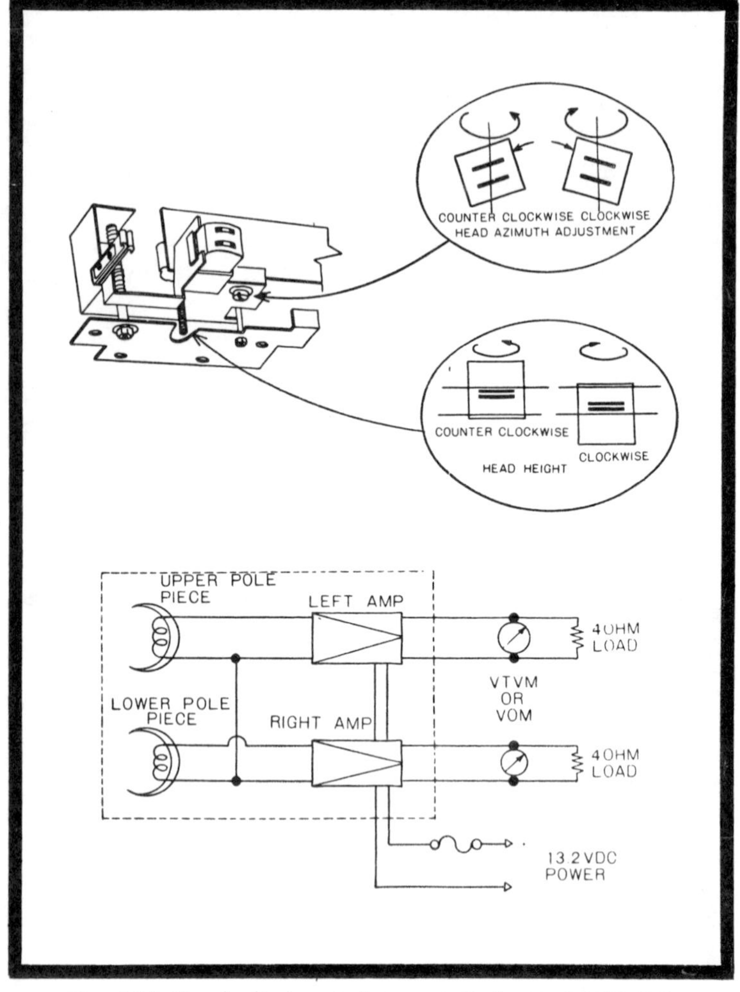

Fig. 4-13. Head adjustment, Automatic Radio Model 57-6ES.

Here are two typical examples of head alignment in auto and home-type tape players. The output of the auto player is monitored with a VTVM or VOM and a load resistor whose value equals the impedance rating of the loudspeaker. The home tape deck output is monitored with a VTVM and oscilloscope.

Auto Tape Player Alignment. The auto tape player in this example is an *Automatic Radio* Model 57-GES. The test setup is pictured in Fig. 4-13.

1. With the cam in the low position, the bottom edge of the pole piece should be in line with the bottom of the tape guide. If out of alignment, adjust using head height adjusting screw.
2. Using the 8 kHz portion of the test tape, set for minimum output, using azimuth adjusting screw.
3. Repeat steps 1 and 2.
4. Using the 1 kHz portion of the tape, check for uniform output on all four channels.
6. It may be necessary to go over all sets of tracks two or three times to achieve perfect alignment.

Home Player Alignment. The home unit for our example is a *Masterwork* Model M-50A. The azimuth adjustment sequence is given in the following steps.

1. Connect scope and VTVM to speaker terminals.
2. Insert tape cartridge (RCA 326).
3. Set volume control to maximum nonclipping output level on scope.
4. Adjust screw 1 (Fig. 4-14) to maximum nonclipping output.

To check crosstalk, the head position should be observed. The setup is basically as described in the azimuth adjustment procedure, but the test tape will be an RCA 327. The sequence is as follows:

1. Set volume control to maximum.

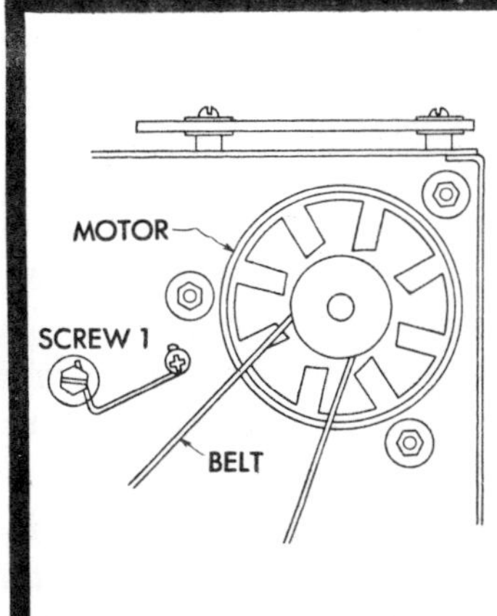

Fig. 4-14. Adjusting head azimuth.

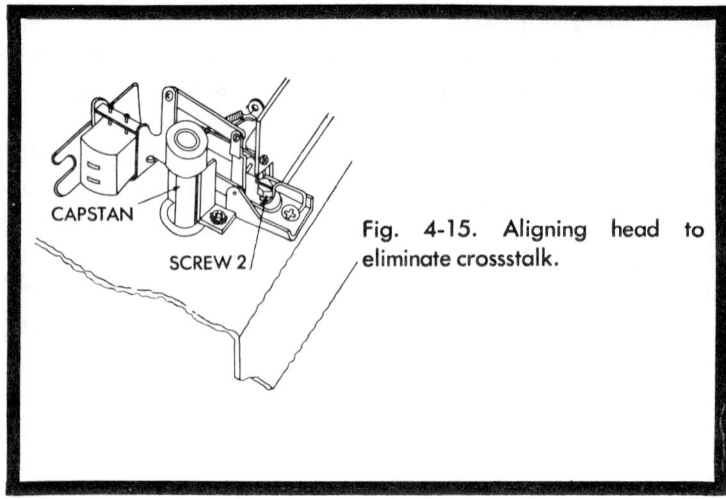

Fig. 4-15. Aligning head to eliminate crossstalk.

2. Adjust screw (Fig. 4-15) for minimum output of undriven channel. Then switch programs and repeat.
3. Change cartridge to RCA 328 and repeat for alternate tape tracks.

For the output level adjustment, leave the equipment connected and insert test cartridge RCA 323. Set volume to maximum, and adjust any internal variable volume-setting resistors for a very slight clipping of the waveform pictured on scope.

Head Adjustments Without the Test Tape

If a test tape is not available, an ordinary tape with a vocal, piano or guitar selections may be used. In this case, head height adjustment is satisfactory when the pauses between words are free of sounds from adjacent tracks. Be sure to check the first (program 1) and last (program 4) positions of the tape head assembly for crosstalk.

If a test tape is not available, the background hiss on any cartridge may be used for azimuth alignment. Units with Dolby noise reduction must have the Dolby switches turned off, of course. All that is necessary here is to adjust for maximum meter deflection (audio volume). A high-pitched sound indicates improper head angle or tilt adjustment. A balance or level meter may also be used as the output indicator for head adjustments.

Improper Adjustments

When the adjustment screws have been turned way out of line or a new tape head has been installed and the tape head does not rotate over all four channels, start at program 1 (when tape head is at the highest point). Generally, at this position excessive crosstalk is readily noted and only one speaker may be functioning. Select a known tape cartridge with music on position one (most cartridges list the number of musical selections of each program channel). Adjust height screw for well balanced music of the same program from both speakers. Rotate balance control to extreme right and left to check for crosstalk interference. Now press the channel selector (manually) to select program 4. Recheck known music on the last track of the tape cartridge. Readjust height screw for maximum output on both channels. Then check each position for crosstalk and proper tape head balance.

Tape Balance Adjustment

Located in some tape players is a separate tape balance control. This control should not be adjusted unless the balance is way off, and then only when a prerecorded tape is available to verify that out-of-balance condition is not a recording anomaly (rather than a playback problem). The partial schematic of Fig. 4-16 shows the relationship of the tape balance control with the conventional left–right balance adjustment. Connect a VTVM or balance meter to both left and right speaker output terminals for this adjustment. Adjust the control for equal signal on left and right output terminals.

Speed Adjustments

A separate motor speed control circuit may be found in many auto stereo tape players. In some models this control is located near the front panel and adjusted with a long screwdriver; the top cover must be removed in other models to gain access to this control. A special speed test cartridge can be used in this speed adjustment. This tape has short clicks at precisely timed intervals. By clocking the time between clicks, speed errors are observable. (These tapes

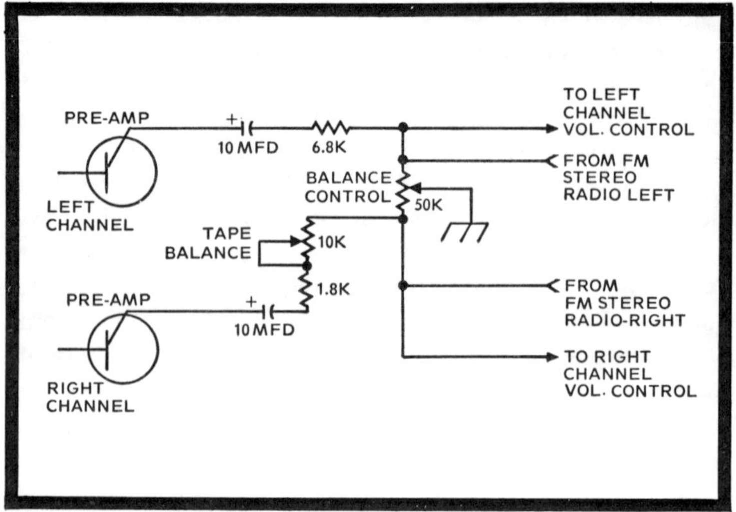

Fig. 4-16. A separate tape balance control is located in some auto tape player models.

carry a voice identification that tells the time to expect between clicks.)

Ratchet Solenoid Indexing Adjustment

Since a great amount of current is involved in ratchet solenoid channel changing, some auto tape players have a ratchet solenoid indexing adjustment. This adjustment moves the inside plunger and solenoid closer or farther apart to acquire continuous program cycling. You will find in some models the solenoid mounting screws are loosened to slide the solenoid back and forth. After correct adjustment the screws are tightened into position. An index screw is found in other models at the rear of the solenoid for ratchet indexing adjustment. Home tape players do not normally require solenoid index adjustment.

Head Tension Adjustment

The head tension adjustment screw is generally located on top and to the rear of the tape head. Do not mistake this adjustment screw for the tape head height adjustment. This tension screw should be adjusted when the tape head assembly has been dismantled or the screw has dropped out.

The tension screw is only found in certain models and should be adjusted according to the manufacturer's specifications.

Flywheel Play Adjustment

In some auto tape players a flywheel adjustment screw is located at the end of the flywheel. Tightening the screw reduces the vertical "play" of the flywheel. The flywheel play in other models can be taken up by loosening screws holding the bottom bracket plate. Adjust plate bracket for no more than 15 mils of end play. Then tighten all screws holding the plate bracket. You may find a lot of end play when the bottom bracket is bowed downward. Just remove the bracket and re-form it.

Cartridge-Holding Tension Adjustment

The cartridge-holding tension pressure is controlled by wire springs, detent spring, and roller. This cartridge pressure is applied against a notch on one side of the cartridge. The tension pressure will vary from 5 to 10 lb for different tape players. When the tape cartridge will not seat properly against the capstan drive, replace springs or spring detent roller assembly. Sometimes you can quickly remedy the situation by cleaning the plastic rollers and metal sides. Oc-

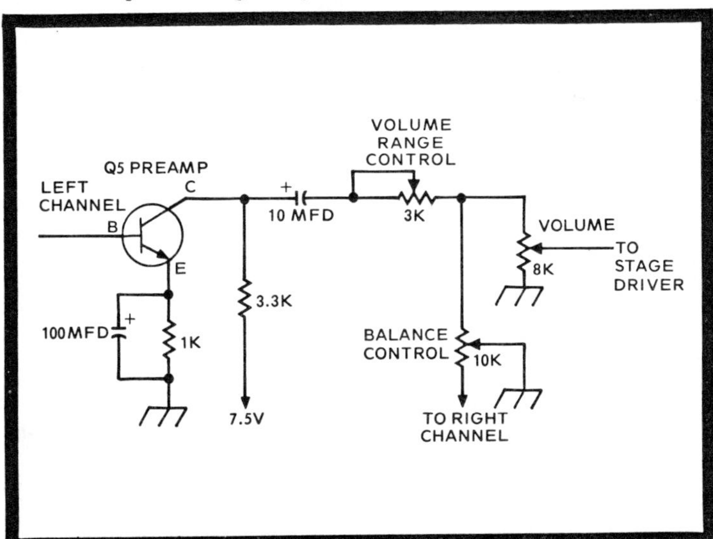

Fig. 4-17. Circuit of tape volume range control in stereo 8 players.

casionally the problems will disappear when a few drops of oil are applied to the plastic roller bearings.

Volume Range Balance

In some stereo amplifier circuits a volume range control is located just ahead of the balance control. Each channel amplifier has a separate control besides the regular volume control (Fig. 4-17). Select a test tape cartridge with equal output of left and right channels. Use a VTVM or balance meter as an indicator. Set the balance control to its center of rotation and adjust each separate control for equal undistorted output.

Cleanup and Lubrication

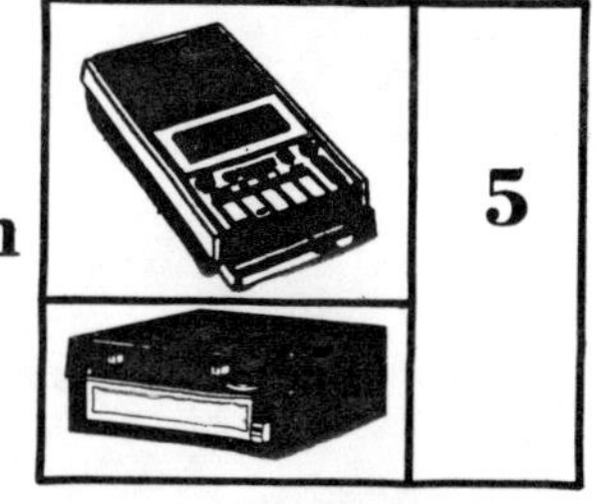

5

Magnetic tapes and players require proper maintenance if optimum results are to be expected. Due to the constant friction between the tape, the tape head, and the drive assembly, you will find an accumulation of oxide. Cassette and cartridge tapes use an oxide on one side and a graphite lubricant on the other. After extended use the tape playback head and the drive capstan will build up a layer of iron oxide from the tape. The layer of oxide upon the tape head prevents the tape from making full contact with the head, which results in a gradual loss of high frequency response and an increased noise level. In many cases accumulated oxides from past tape playings will cause a loss of sound and excessive distortion. The oxide deposit on the capstan can cause slippage (wow), which is apt to be taken for more serious mechanical drive problems. If nothing else, built-up head oxides will cause serious high-frequency loss and audio level problems. The tape head should be cleaned periodically—and especially when it comes in for repairs. (Cleaning involves nothing more than swabbing the head with alcohol or a specially compounded solution, as shown in the photo of Fig. 5-1.)

Wow and slow speed problems can be caused by improper lubrication or oil upon the moving drive surfaces. Frozen or dry capstan bearings produce dragging, freezing, or slow speeds. A dry or worn motor bearing may cause drag or wow conditions. Oil or residue upon the drive belt will produce inconstant speeds. Sometimes a good cleaning and lubrication may solve the tape player's problem. So let's see how easy it is to clean and lubricate the tape player.

CLEANING SOLVENTS

There are many commercial cleaning solvents available. They can be applied with cotton swab or with a spray can.

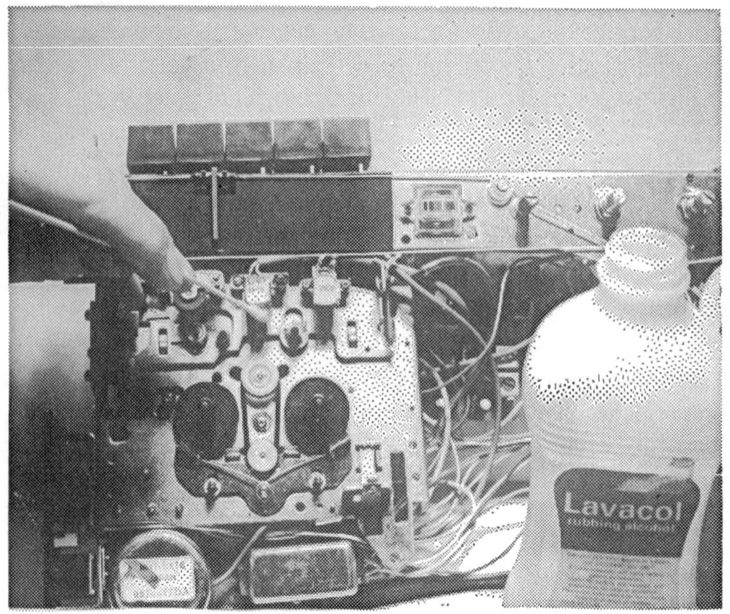

Fig. 5-1. Cleaning tape heads on a cassette player.

These commercial cleaners are all good but plain alcohol found in most drug stores is inexpensive and does the job. Purchase a good grade of isopropyl alcohol. Be careful not to let the solution run down the capstan into drive bearings. Keep alcohol off of all plastic drive surfaces and plastic front pieces. Sometimes alcohol rag wiped against certain plastic fronts will tarnish or dull the surface. Some of the better commercial solutions have a dye that adheres tenaciously to plastic surfaces.

TYPES OF LUBRICATION

Most manufacturers recommend that lubrication is needed only after each 1000 hours of use or every 6 months of operation. In many cases bearings do not need lubrication unless moving parts tend to bind or are replaced. When necessary to lubricate, first remove all of the excess or old lubrication around the component. Be sure to keep the pressure roller, tape head, capstan, belt, and idler completely uncontaminated—they must be absolutely free of any oil residue. Be careful not to apply too much oil or grease—do not overlubricate. More often than not, only a drop is needed. Overlubrication is virtually certain to be detrimental to proper tape operation. The ideal lubricant is a very light high-grade machine oil, such as the popular *3-in-1* brand available at most drug and five-and-dime stores.

Light oil should be used on bearings; grease is for sliding surfaces. Only a drop of oil is needed upon roller, idler, motor, and capstan bearings. When applying lubricant from a spray can, never spray directly onto the bearing. Spray the oil onto the end of a swab stick or pencil, and then apply oil from the swab to the bearing. Try another approach if you like, but be forewarned that oil might fly all over the belt and capstan drive surfaces, causing slippage that can be tedious to correct.

PERIODIC CLEANUP

The tape head should be cleaned periodically with alcohol-moistened swab or tape cartridge cleaner. The playback head can be cleaned, without disassembling the player, by reaching through the tape cartridge door and lightly wiping the head with the moistened swab. Also, the capstan can be cleaned from the front opening. Depress the momentary-contact cartridge switch with a pencil or probe and apply the moistened swab to the capstan as it rotates. Owners tend to forget the weekly head cleanup chore, and the result is deterioration of sound quality and speed problems.

Cleaning the Tape Head

In most cases the ordinary *Q-Tip* swab won't reach the head when you're trying to service a dash-mounted automotive cartridge player. (This applies equally to many auto cassette models—Sanyo, for example.) But before you start removing the player, you should know that many electronics part houses carry swabs of double length—for just this purpose. And if there are none available, you might think about buying a hemostat-type seizing tool. This tool is similar to a pair of long-nose pliers but it's considerably more delicate. Doctors use it to clamp off veins and arteries during surgical operations. Unlike the long-nose pliers, the hemostat may be locked in a closed-jaw position; this allows the tool

Fig. 5-2. Cleaning tape head on the cartridge tape player.

and swab to become one long assembly that will easily reach into the tape receptacle as far as you need to go.

For thorough cleanings when you're servicing an unmounted unit, remove the cover so that you can better see the oxides you're missing (Fig. 5-2). With the covers removed you can do a more complete cleanup job on the bench. Be on the lookout for telltale oxide particles everywhere—even *under* the main deck of the chassis. The unit pictured in Fig. 5-3 had clean heads but oxide buildup on capstan and guide assemblies caused erratic operation.

Alcohol won't handle the really tough buildup problems. And if you don't have one of the supereffective solutions (like Robins TX-20), you may have to use a scraper. Take a plastic rod or wooden dowel and grind down both sides like a screwdriver blade. Generally, light pressure with such a tool will remove the packed oxide without scratching the metal surface. *Never* use a knife or screwdriver blade. You may scratch and ruin the tape head. Also, metal tools should not touch or be near the tape head since they may become magnetized and then magnetize the head. When working with metal tools around the tape head, make certain they've been demagnetized.

Cleaning the Capstan—Flywheel

For proper cleaning of the capstan and flywheel assembly, they should be removed from the player. To remove a flywheel from a cartridge player, remove two small metal screws holding the flywheel support bracket (Fig. 5-4). In some models, the capstan—flywheel assembly may be held in position with metal brackets or a *C*-washer (at the capstan end). Others use a metal pin that goes through the metal housing to hold the capstan in position. If you don't remove the flywheel, the player will likely be back in for service in a very short time.

Use a cotton swab or pipe cleaner soaked in alcohol to remove iron oxide, graphite, and grease from the capstan bearings (Fig. 5-5). In commercial cleaning kits a bristled cleanout brush is included for these bearings. Be very careful not to misplace any of the rubber or nylon bearing washers. Some capstan bearings have a felt oil washer that should be cleaned or replaced periodically.

Clean all surfaces of the flywheel and capstan shaft (Fig. 5-6). Notice if the capstan surface is scored, rough, or dirty. A defective or scored capstan will pull tape from the cartridge

Fig. 5-3. Cleaning tape guide assemblies and under the tape head.

Fig. 5-4. Cleaning belt and flywheel in a stereo 8 player.

Fig. 5-5. Cleaning out capstan drive bearing with cotton swab or thin brush.

and will make the tape "ride up" out of its guides. Clean off the belt surface of flywheel with alcohol and lint-free cloth or paper towel. In some cases small particles of rubber (from drive belt) will still cling to the flywheel and may have to be removed with a plastic dowel.

Always allow several minutes for the head-cleaning solution to dry completely before inserting a tape or tunning the motor drive belt. And don't forget to lubricate all bearings before replacing capstan–flywheel.

Motor Belt and Drive Pulley

When a loud rush is heard with volume up and no tape movement, suspect a thrown-off drive belt, caused by age, loosening, or operator abuse. A frozen capstan–flywheel will let the motor spin inside the drive belt. If the drive belt is too loose, wow and flutter conditions will result. Oil on the drive belt will cause unstable operation. Replace the drive belt if the existing belt won't stay in position. See if the motor pulley, drive belt, and flywheel are in line.

Check the drive belt for cracks or worn areas. See if the inside section of the drive belt has a gray or shiny appearance—this indicates the drive belt has been slipping. Small rubber particles under belt and on motor pulley indicate slip-

Fig. 5-6. Cleaning capstan—flywheel. Notice area of wear on capstan shaft.

page, too. Clean the belt surface with a solution compatible with rubber and do not stretch the small belt while cleaning.

Most motor pulleys are secured to the small motor shaft with a setscrew (a loose motor pulley will throw the drive belt and produce wow conditions). If, after cleaning the motor pulley, rubber particles still cling to the surface, remove with a plastic dowel.

Square belts are often particularly troublesome. When twisted, they produce wow conditions. When replacing, be sure the small drive belt is going in the proper direction and around the right pulleys; otherwise, you will find the takeup hub going in the wrong direction, causing tape spillage and fouling. When a separate drive belt is used for fast-forward and rewind positions, a dirty or oily belt will produce sluggish speeds. It's best to have a good stock of various drive belts on hand—especially if you do warranty repairs for a certain brand of players.

Motor Cleanup

When the motor shaft is frozen or sluggish, the motor should be taken apart and cleaned. Details on removing and dismantling the small drive motor is given in Chapter 3. Replace the entire motor assembly when the bearings seem to chatter and are excessively noisy. A drop of motor oil may cure a noisy motor bearing, but this is generally a temporary repair.

When the motor shaft is frozen, dismantle the motor and remove the end bearings. Clean out the bearings and motor shaft with alcohol. Check the bearings for excessive wear. In some instances these motor shafts are extremely worn and cannot be replaced. In this case you should replace the entire motor assembly. Most motors have sealed bearings and do not require lubrication. If not, lubricate the bearings and shaft with a light machine oil. Be careful not to let oil run down upon small brushes to cause excessive arcing on the armature surface.

Ratchet Assembly Cleanup

A gummed-up ratchet assembly in a cartridge player may cause sluggish and erratic operation in changing channels. If the player is less than a year old, the ratchet assembly should not be oil or cleaned. When repairing a defective ratchet assembly, clean and lubricate the moving parts. Wash off old grease from ratchet channel-change actuator, and remove oil and grease from head-shifting cam. When the grease is difficult to remove from the shifting cam, remove solenoid plunger and detach from ratchet assembly. Wipe off oil or graphite grease from solenoid plunger and inside solenoid sleeve.

Apply a thin coat of grease to head-shifting cam. Lubricate ratchet actuator with light oil. Most solenoid plungers are not lubricated, while others have a light coat of graphite grease. If in doubt, do not grease at all. A drop of oil on the head changing lever will suffice.

Cleanup of the Undercarriage

Tape oxides will be found on the pan under the tape head assembly. Always clean off all grease and oxide from chassis parts and under carriage pan (Fig. 5-7). If the oxide is not removed from under tape head assembly, it falls into capstan bearings. This will bind the bearings and cause speed problems. A "wash" spray with a plastic spout is ideal to re-

Fig. 5-7. Excessive oxide on stereo 8 tape player.

move grime from underneath. Tv washout lube will also do a good job. Wipe up excess spray and grease with a paper towel. If a cotton swab is used, keep it away from the tape head surface. Also, clean out the cartridge compartment and wipe off the plastic rollers. A drop of oil on the plastic roller bearing will let the cartridge glide in and out smoothly.

Final Cleanup

Don't forget to clean up the case and control panel after the tape player has been repaired and tested. You can clean off dirt and grease with a window spray. Spray into small crevices and over the entire case. Use a small stiff-bristled brush to remove dust from difficult places. Then wipe off with a rag or paper towel. It's quite surprising how window spray will make chrome surfaces and the entire player shine like new.

Remove small knobs and control levers and spray their panel areas with window cleaner. Wipe foam off with a paper towel. Use a sharp-pointed tool to get into liner and crevices of the control knobs. Clean out all grime and dirt. If necessary, spray once again and wipe dry.

When the metal cabinet of auto stereo tape players is scratched and scarred, a coat of paint will brighten up the unit. First, clean up the metal cabinet. Mask the front control panel. Select the right color paint and spray a coat on top and sides. Let dry, then spray the underside. It takes only a few minutes to spray a tape player, and final results are quite rewarding. All wooden cabinets on home tape players should be cleaned and polished with lemon oil or furniture polish.

LUBRICATION

A lubrication procedure from a typical manufacturer's service manual is given below:

All mechanical parts are factory lubricated and should require no additional attention unless moving parts are binding or are replaced. Whenever lubricant is required, a very small amount of the specified lubricant should be applied. These instructions were given of a *Realistic* (Allied/Radio Shack) home stereo tape unit, Model TR-284.

1. Apply a small amount of Launa 40 oil upon the bearing area of the flywheel bearing.
2. Apply a small amount of Beacon 2 oil upon the bearing ball between thrust bearing and flywheel shaft.

CAUTION: Tape contact area of the capstan must be kept free from contamination with lubricant.

NOTE: Launa 40 oil is a product of Nippon Oil Co. Ltd. Beacon 2 oil is a product of Esso Standard Oil Co.

Lubrication of Capstan—Flywheel Assembly

A dry or noisy capstan drive bearing should be removed for cleaning and lubrication. Bright-and-shiny bearings may be pretty to look at but they indicate a lack of lubricant. After cleaning the capstan—flywheel, don't forget to lubricate bearing area before replacing. Some capstan—flywheel assemblies have three separate areas to clean and lubricate. Lubricate top and middle areas with machine oil and the bottom support area with a light grease.

Check the capstan muin bearing for excessive play before replacing the support plate. (A worn capstan bearing will produce wow and erratic speeds.) In some models there is a nylon washer that tends to become dry. When this occurs, the washer works it way upwards, causing wow. If the bearing is excessively worn replace entire capstan—flywheel assembly.

Switch Cleaning and Lubrication

The cartridge-triggered power switch has contacts that may be cleaned with solvent or TV-tuner lube. If the points are accessible, insert a piece of cardboard between them to burnish the points (by holding the points together and moving cardboard back and forth). A section of a tube carton or a business card works nicely here. In some models, the cartridge switch is mounted behind the capstan drive assembly and is triggered with a thin metal sleeve. The cartridge pushes against the sleeve and activates a mechanical switch arm. Sometimes this switch will lock in the *on* position and the tape player will not turn off. Remove the switch assembly and straighten the metal sleeve. Clean out all grime and oxide grease. Place a thin coat of light grease or oil over the sleeve. Replace switch and double-check with a fresh cartridge.

Lubrication of Drive Wheel and Motor

After the drive wheels have been cleaned in a cassette player, a drop of oil upon the bearing should suffice. Wipe off any excess oil which may throw out and accumulate on the rubber tires. If the rubber tire appears to be cracked or old, replace the drive wheel.

Most motors are prelubricated for at least 1000 hours of operation. If the motor appears to have a noisy or dry bearing, place a drop or two of oil on each end bearing. In case the motor shaft is frozen the motor should be dismantled and cleaned before lubrication. Be sure and check the shaft to make sure it is true (not bent). Oil very lightly.

Lubricating Tape Head

A lubricating solvent can be applied to tape head and guide assemblies for smoother tape operation. Besides insuring smoother operation, it prevents oxide dust from packing on the tape head and is completely safe to use. Unfortunately, the lubricating qualities are short-lived; the procedure should be repeated often.

Excessive Lubrication

You'll see tape players from time to time in which the owner has "oiled everything." Excessive lubrication can be highly detrimental to proper operation. Wow and slow-speed problems are produced by oil on belts and drive wheels. When excessive lubrication is the case, wash and clean up all moving surfaces with cleaning fluid. You can clean out oily surfaces with tuner wash. Then lubricate according to the manufacturer's service manual or follow instructions below.

Complete Cassette Player Lubrication

The principal components of a cassette player that may be lubricated with machine oil are the pressure roller arm assembly, the head plate assembly, and the turntable assembly (Fig. 5-8). For a thorough lubrication, also oil these functional elements:

- Switch function lever
- Flywheel—capstan bearing
- Button holder lever
- Counter holder bearing
- Takeup arm pivot
- Takeup pulley bearing
- Pulley reel wheel bearing
- Fast-forward pulley bearing
- Any other bearing surfaces
- Pulley flywheel drive bearing

The following parts of a cassette player may be lubricated with a light coat of grease:

- Pushbutton slides
- Button flywheel bearing support
- Sliding head assembly
- Slide shuttle assembly
- Record lever interlock arm
- Sliding lever arm
- Any other metal sliding surfaces

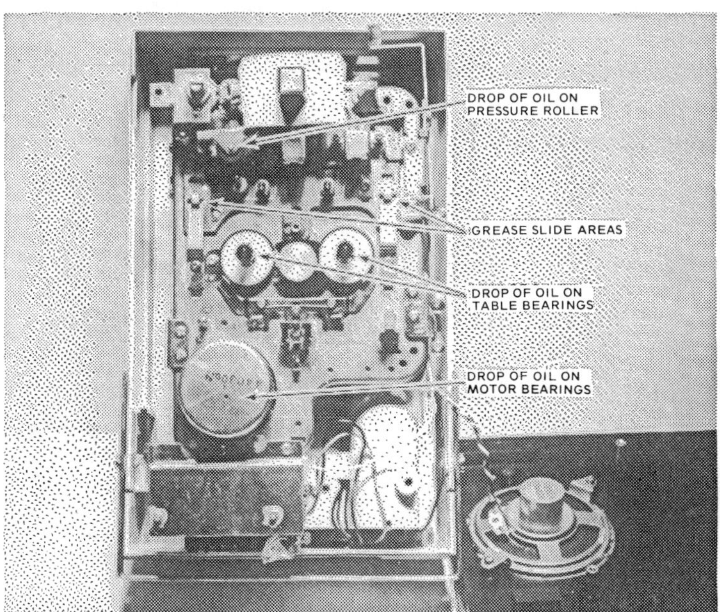

Fig. 5-8. Areas to be lubricated in a cassette player.

Fig. 5-9. Areas to be lubricated in a stereo 8 home tape player.

Complete Cartridge Player Lubrication

The following components of a cartridge player may be lubricated with machine oil (refer to Fig. 5-9).

- Capstan—flywheel bearings
- Motor bearings
- Head elevation arm holder
- Head elevation arm bearing
- Cartridge guide rollers
- Cartridge pressure roller
- Switch assembly
- Track changing lever

Apply light coat of grease to the head cam assembly, capstan bottom support bearing, cam ratchet plate, and the ratchet guide.

Speed Problems

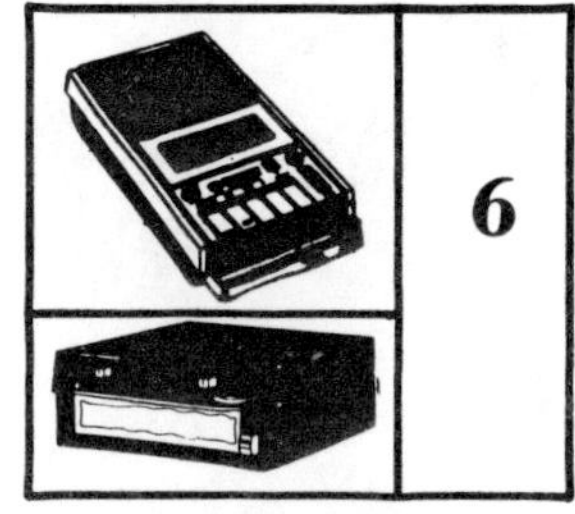

6

Speed problems will account for 75–80% of the troubles found in cassette and cartridge tape players. Slow speeds are mostly caused by mechanical problems, while fast speed may be produced by mechanical or electronic defects. Figure 6-1 shows the most common sources of "sluggish speed" problems. Wow and flutter conditions are the results of uneven tape speed. Although the cassette player has more working components to produce speed problems, both players are similar with respect to servicing.

SPEED PROBLEMS IN CASSETTE PLAYERS

To avoid possible confusicn in servicing tape players, the two types are dealt with in separate parts of this chapter. The following problems represent those typically found in auto and home-type cassette players.

No Tape Motion

This problem may be caused by weak or defective batteries. Weak batteries will result in slow speeds and distorted sound. Check to see if the cassette motor will rotate in ac operation. If the motor still does not function, suspect poor battery connections. Old batteries left in tape players will corrode and attack the spring contact points. Clean and polish the battery holder contacts with a pocket knife, file, or sandpaper.

If the motor does not rotate in ac operation, suspect a defective power switch (Fig. 6-2). In some models a leaf-type switch is used; it may be bent out of position or have dirty contacts. Check for continuity of the switch with a VOM. Be sure and check the AC/BATTERY switch found in some models, in addition to the ON/OFF switch. Check the ac power supply if

Fig. 6-1. Slow speed areas in the cassette player.

Fig. 6-2. Defective power switch in a cassette player. Some smaller players have a leaf-type switch.

the player operates on batteries and not when household power is used as the source voltage. (A typical power supply circuit is shown in Fig. 6-3.)

A defective motor may cause no or intermittent tape motion. The small motor may have bad brushes, an open winding, or frozen bearings. Remove the belt and spin the motor pulley. If the motor shaft is free, check for correct voltage at the motor terminals. You may find two motors in some cassette players (Fig. 6-4). One motor is used in the play–record mode while the other motor operates the fast-forward and rewind operations.

Capstan Does Not Revolve

First check to see if the motor rotates—the drive belt may be off, broken, or loose. See if the motor pulley is rotating inside the drive belt. Drive belts found in cassette players are quite small and are generally loose compared to cartridge motor belts. If the motor and belt appear normal, suspect a dry or frozen capstan bearing. Remove capstan–flywheel bracket assembly and inspect capstan bearing (Fig. 6-5). A good cleanup and lubrication may solve most frozen and sluggish capstan–flywheel problems. Also, check for a binding pinch roller.

Slow Cassette Tape Speeds

When the tape speed is slow in battery operation replace all of the small batteries. If the tape speed is slow in both ac

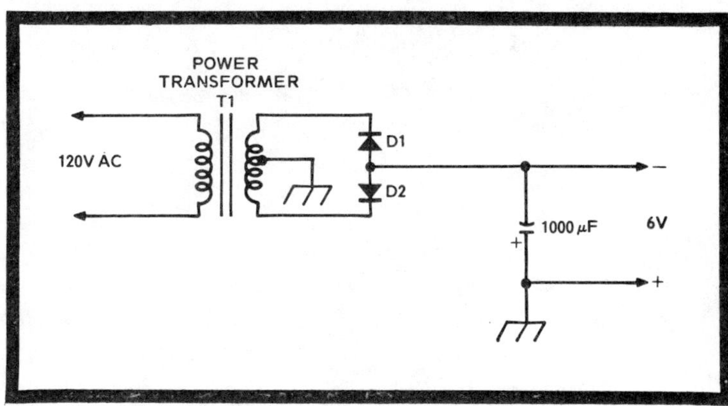

Fig. 6-3. Typical ac power supply for cassette players.

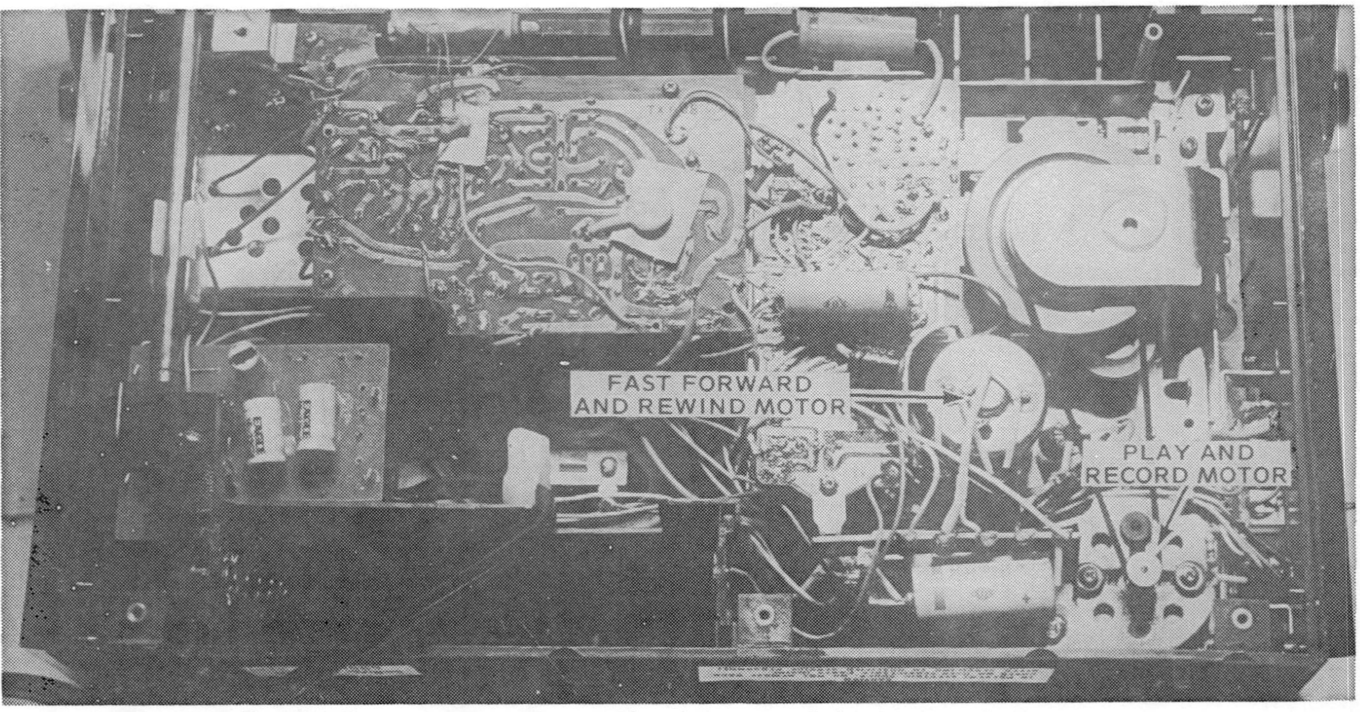

Fig. 6-4. Two separate motors are found in some of the cassette players.

Fig. 6-5. Remove the capstan—flywheel to inspect capstan bearing. Notice how small the tape drive shaft is in diameter.

Fig. 6-6. Check for worn or deformed pinch roller in the cassette player.

and battery operation, suspect a loose or slipping motor drive belt. Clean the belt with alcohol or replace it. Remove the capstan–flywheel and clean the shaft. Properly lubricate and replace. Check the pinch roller for dry bearings. With capstan drive disengaged the pinch roller should spin freely. Proper tension of the belt and pinch roller should be followed according to the manufacturer's specifications.

Excessive Wow and Flutter

A damaged or loose motor belt may cause wow conditions. Check the square motor belt to see if it is twisted between motor pulley and capstan–flywheel. Clean off dirt and oil deposits upon the motor belt. Check the flywheel for dry or frozen bearings. If excessively worn, replace the flywheel–capstan assembly. Oil on the flywheel will almost certainly cause wow. Check for a faulty flywheel bracket; adjust accordingly. Wipe off all moving parts with alcohol and a moistened cotton swab.

Suspect a dirty, deformed, or oily pinch roller for flutter conditions, particularly when there appears to be some regularity to the flutter. A deformed or worn pinch roller should be replaced (Fig. 6-6). Insufficient pressure of pinch roller against tape may cause wow conditions. Adjust the pinch roller spring for correct pressure according to the manufacturer's specification. If necessary, replace the pinch roller spring.

Other common causes of flutter (and wow) are insufficient tension on the takeup reel, improper winding of tape in cassette, and badly repaired tape breaks. With stubborn flutter problems, use a new cassette that has been previously recorded on a known-good machine. (Be sure to take up tape slack manually before checking.)

Improper cassette pad pressure may cause wow and flutter. The pad pressure should be adjusted to the manufacturer's specification. If the takeup torque is too high replace the takeup mechanism.

Fast Tape Speeds

When the tape momentarily rotates faster or skips, suspect insufficient pinch roller pressure. Measure the pinch roller pressure. If insufficient, increase it to the specified pressure according to the manufacturer. In some cassette players, simply changing the position of a spring to a more distant anchor point will take up the pinch roller pressure. If the takeup torque is excessive, the tape will be squeezed out of alignment by the capstan and pinch roller—replace the drive pin assembly. In case the capstan and pinch roller are not parallel with each other, a change of speed will be noted. Simply adjust the pinch roller mount. Also, suspect a defective motor governor or motor for fast tape speeds.

Sluggish Fast-Forward

When *fast-forward* operation appears to become sluggish suspect insufficient pressure of intermediate idler pulley. Check for loose or out-of-position roller spring. If fast forward torque is insufficient, clean oil from the turntable on the fast-forward roller. See if the brake plate still touches the spindle (resulting in sluggish forward speeds). A defective cassette can cause sluggish operation, of course; don't overlook this possibility.

Fig. 6-7. In fast-forward position, check for loose or oily belt that can cause sluggish and erratic speeds.

torque is insufficient, clean oil from the turntable on the fast-forward roller. See if the brake plate still touches the spindle (resulting in sluggish forward speeds). A defective cassette can cause sluggish operation, of course; don't overlook this possibility.

Sluggish or Slow Rewind

When the rewind speed is slow check for slipping at the roller and drive belt. Clean off all driving surfaces. Check for insufficient engagement of right and left idler assemblies. Check for correct pressure of the rewinding roller. Adjust the takeup torque according to manufacturer's specification. Suspect a brake plate that does not release. Check for a defective hub of the cassette, too; a warped cassette will produce slow and sluggish operation.

Pressure Roller

The pressure roller exerts pressure against the tape which is between the roller and capstan drive shaft. Pressure roller tension is more critical than normal since the drive area is quite thin on the capstan drive assembly. If the tension is too high or too low, wow and flutter result. A dry, worn, or deformed pressure roller will cause the same problems. When not engaged the pressure roller will spin freely. If necessary to adjust, carefully re-form or replace the pressure roller. Don't forget to also check for correct pressure of the tension spring.

Gradual Slowdown

When the tape slows down after the player has operated for several minutes, suspect a binding or dry capstan–flywheel assembly. Remove the capstan; clean and lubricate bearing areas. Check the motor for dry or warm bearings. Also, the motor may be defective. Suspect a defective cassette and compare with a new one.

Takeup Tension

In case of absence of movement or erratic takeup, clean the spindle and pulley drive shaft. Tension on the takeup hub is determined by the friction between the takeup tension pulley and takeup drive shaft. Normally no adjustment is necessary, but if the tape fails to take up properly, check the takeup arm assembly spring. If necessary, carefully re-form or, if needed, replace the spring. Double-check the suspect cassette with a new one.

Excessive Mechanical Noise

First, determine in what area the noise is originating. The noise may be caused by a defective capstan–flywheel, pulleys, motor, or cassette. Remove the cassette and see if noise is still present. If not, insert a new cassette. In case the noise is still present remove drive belt and check motor. Dry motor and capstan–flywheel bearings will produce squeaks, chattering, or scraping noises. On one high-quality deck, a rarucous scraping noise was traced to the main drive motor. Disassembly showed that one of the armature fan blades had worked loose from the armature. A good glue remedied the problem.

Checking for Wow or Slow Speeds

Slow speeds in the cassette player can be checked with a test cassette or a recorded cassette containing piano or guitar music. Notice the pitch of tone or music for slow and wow conditions. Try a new cassette if one is suspected of erratic speeds. After servicing the cassette player, check for correct speed and azimuth adjustment (Chapter 4).

SPEED PROBLEMS IN CARTRIDGE PLAYERS

The following problems are typical of those found in 8-track cartridge players.

No Pilot Light or Tape Motion

When the tape player fails to play and no power appears to be getting to the machine, suspect a blown fuse or open cartridge switch. Check for dc voltage between one terminal of the cartridge switch and ground. If no voltage at this point, suspect a defective plug or broken wire connection. Burned or open PC board connections can cause power loss, too.

If the pilot light is on and you can hear amplifier hiss, suspect a broken belt or defective motor. A broken or loose motor belt will let the motor spin freely. Frozen capstan drive and motor shaft will result in no tape motion. Check for a loose pulley setscrew on the drive motor. In case the motor does not rotate, suspect a defective motor or absence of dc voltage to the motor. Check for proper dc voltage at the motor terminals.

Slow Tape Speed

A cracked or worn motor belt can produce slow-speed problems. To determine if there is belt slippage, check to see if capstan or flywheel has a shiny appearance. To check for a stretched belt, place one finger on the motor pulley and turn the capstan pulley. If the belt turns easily over the motor pulley, the belt is too loose. A stretched belt will not stay in position and flies off the pulley. Oil on the belt surface produces slow and erratic speeds (Fig. 6-8). It's best to replace the motor belt when the unit is brought in for service.

Besides the drive belt, check the capstan drive and motor for slow-speed problems. A dry or partially frozen capstan drive shaft will slow the tape motion. Spin the flywheel with one finger; it should spin for several seconds. A defective motor with a flat armature or dry bearings will cause slow tape speeds. Electronic speed circuits and defective mechanical governor inside the motor can produce slow tape motion. A partial schematic showing a typical speed regulator circuit is presented in Fig. 6-9.

Erratic Speeds

Erratic or intermittent speeds can be caused by a sticky mechanical governor or electronic speed regulator circuit. First, check for oil spots on motor belt and pulley. If the trouble persists, remove belt and check motor speed. A defective motor may change speeds without a load, or it may not start in one position. Suspect a flat armature or worn armature brushes. Next, check the electronic speed circuit for intermittent speed. Intermittent speed problems may be traced to poor board connections in the speed control circuits and poor motor connections. Check correct seating of the cartridge; a binding or off-center cartridge will produce erratic speeds. Move the cartridge and see if the player speeds up.

Wow and Flutter

A loose or worn drive belt can cause wow. Check the drive belt for grease or an oily surface. Suspect oil on the

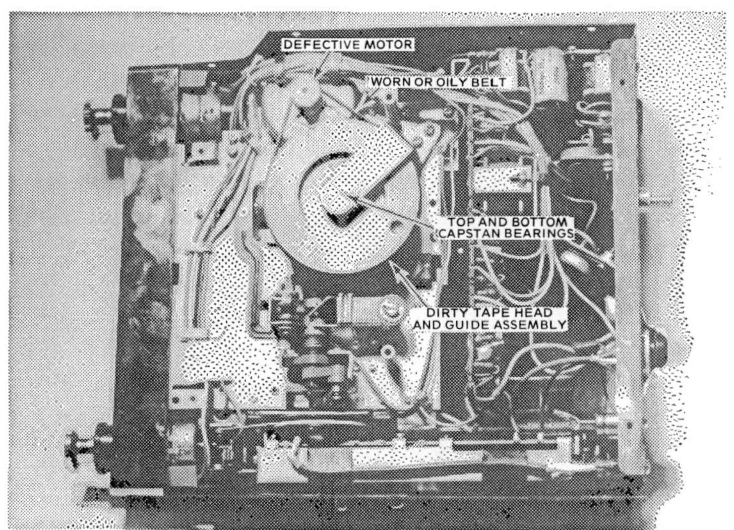

Fig. 6-8. Slow speed areas in the stereo 8 players.

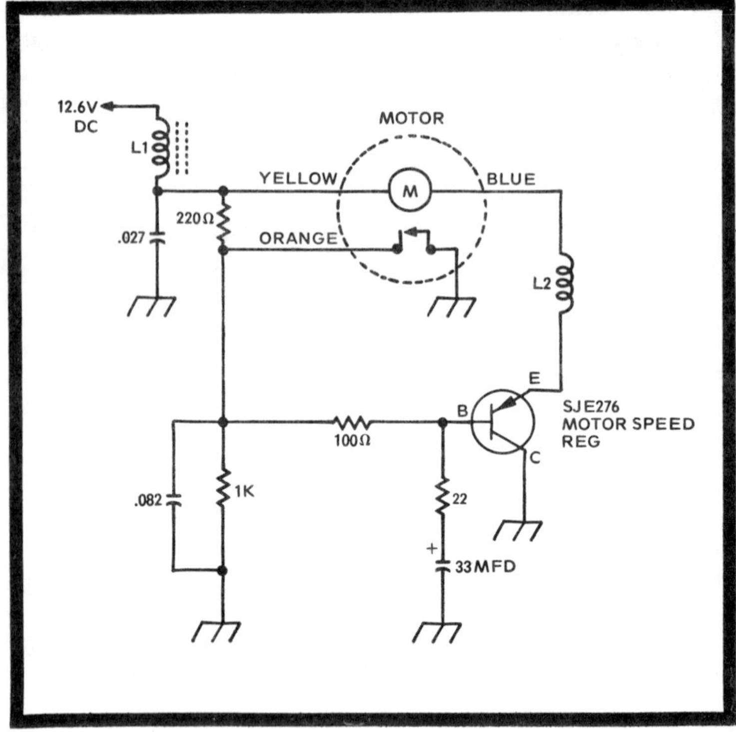

Fig. 6-9. Single-transistor speed regulator.

flywheel and motor pulley. See if the drive belt is riding high on the motor pulley. Inspect the motor pulley for small pieces of rubber under drive belt. Typical trouble spots are shown in Fig. 6-10. You may find these small rubber particles difficult to remove; use a plastic rod or wooden dowel.

Excessive wow and flutter can be caused by a dry capstan bearing, out-of-round or dirty capstan shaft, or dirty, oily, or worn flywheel. A worn capstan shaft and bearing results from improper lubrication or oxide falling into the bearings. In most cases, wow and flutter can be eliminated with proper cleanup and lubrication.

Try a new cartridge when you suspect the trouble is not traceable to the player. But remember, improper pressure of the spring board that holds the cartridge in position will not let it seat against the capstan drive—and this can be a particularly troublesome source of wow, flutter, or erratic operation.

Fast Tape Speed

First, determine whether the tape is actually running too fast or you have a wow condition. Use a test tape or a tape with known-good piano or guitar music. Slow tape speeds are easy to identify. Wow and fast-speed problems take a little longer. Wow is easiest to spot with a prerecorded tape containing an oscillator note of a single frequency—say 1000 Hz.

Check to see if the fast-speed problem is related to mechanical or electronic speed-control circuits. Most fast-speed problems are traced to the electronic speed-control circuits.

Dead Motor

When the drive motor will not rotate suspect no voltage to the motor or a defective motor. If the amplifier has a hissing sound and the motor does not rotate, simply remove the drive belt. Twirl the motor pulley and see if the motor starts to rotate. If the motor takes off suspect a flat armature or bad brushes. A good motor pulley will rotate but won't spin freely like the capstan drive when it's turned by hand. If the motor will not turn, look for dry or frozen bearings (Fig. 6-11).

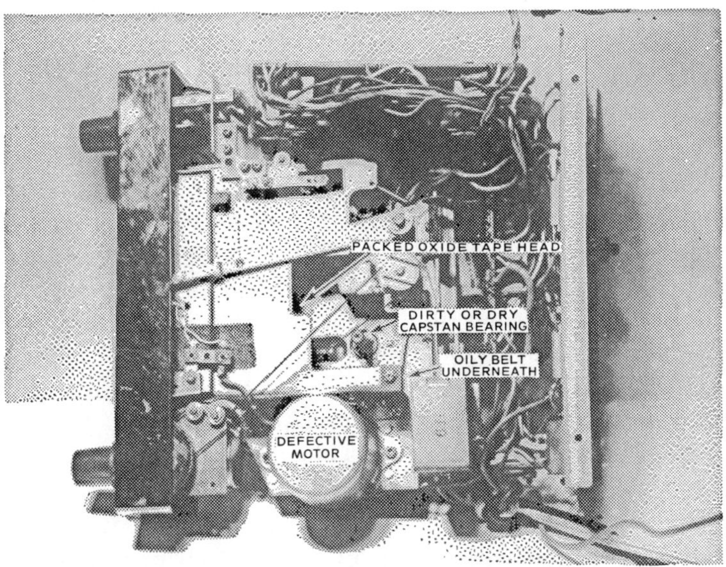

Fig. 6-10. Wow areas found in the stereo 8 players.

Fig. 6-11. Replacing a defective motor in a stereo 8 tape player.

In case the motor will not move at all, measure the motor voltage. The applied voltage should be no less than 12 volts. No or low voltage can be traced to poor wiring connections or defective speed-regulator circuits. Check the motor cartridge switch for dirty, worn, or bent contacts. In some models, two separate switches are activated when the cartridge is inserted.

Motor Runs Slow

Suspect a defective motor if the motor still runs slow after a good cleanup and lubrication. Tap the end of the motor assembly with a screwdriver to see if it changes speeds. If so, the mechanical governor inside the motor is defective. Generally, the mechanical switch points are dirty and pitted. Sometimes these small motors can be repaired by removing the end cover and armature. Clean the switch points with a piece of postcard paper. Generally, this repair is only temporary. The best method is to replace the defective motor with the exact part number.

Gradual Slowdown

When the tape slows down after a half-hour or so, suspect a heat related problem in the motor or capstan drive assembly. Remove the flywheel–capstan drive. Clean off drive shaft and bearings with alcohol. Lubricate and replace. Check the motor bearings. In the newer players, most motor bearings are sealed and do not need oiling; others become dry and noisy. Feel the end bearings and motor shell for overheating. If the motor runs unusually warm, the thermal expansion of the metal parts could be the trouble source. If bearing replacement won't cure it, replace the motor.

Belt Won't Stay Put

If the drive belt will not stay in position, the belt is probably too loose. Check for dry, binding bearings at motor pulley and flywheel assembly. If the motor pulley and flywheel assembly are not in line, the drive belt will tend to work up and fly off. Also, check the position of the belt guide assembly. Some models have a special guide to hold the belt in line with the flywheel assembly. See if the motor pulley is loose on the motor shaft and does not line up with the capstan–flywheel.

Replace a stretched or loose motor belt with the original part number. If not readily available, belt repair kits may solve the problem. These are readily available (Fig. 6-12), and are worth the modest investment.

First, check the original belt for worn and cracked surfaces. In case the belt is in good shape except for stretching, cut out a quarter-inch section. Clean off both ends of the rubber belt. Be sure the ends are square. Use the special cement provided in the kit. Use just enough to cover rubber ends. Now, lay the cement ends together upon a smooth metal or glass surface. Push the two ends together with fingernails of both thumbs—place extra pressure on the spliced joint. Hold the splice in position for about 60 seconds. Now try to pull the bond apart. The cemented splice won't break if made according to directions.

To construct a new belt, select the correct width and size. Use the old belt for correct length. Now pull on the new belt material and see if it will stretch out of shape. If the belt material is now longer than the original, cut off to required length. Clean ends and bond the belt with cement.

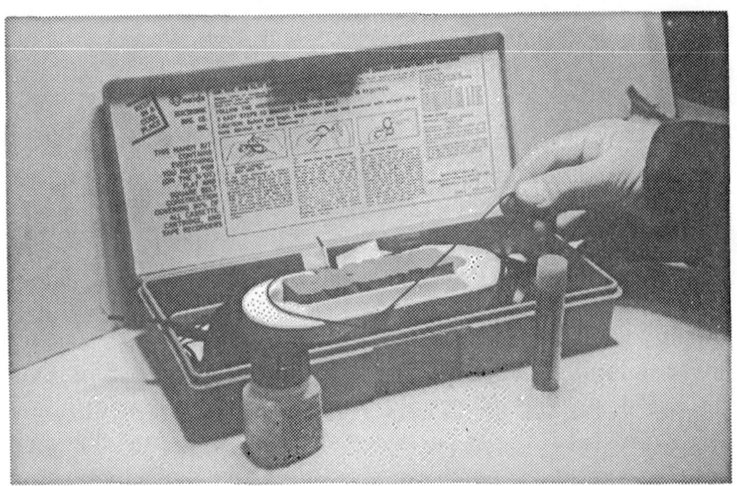

Fig. 6-12. A flat belt repair kit.

Scraping Noises

When the complaint is a scraping noise you can bet the flywheel is scraping against the metal guard or bottom pan (Fig. 6-13). Each time the flywheel revolves, a section will touch against the metal guard. Remove the motor drive belt and spin the flywheel. Make sure the tape player is lying in its mounted position. Now remove the metal guard and capstan–flysheel.

The metal flywheel works down against the capstan drive shaft. In most models, the metal flywheel is molded around the drive shaft and is not secured with a metal pin or setscrews. If a replacement flywheel assembly is not available, do not try to pin the flywheel to capstan drive shaft—the drive shaft material is too hard. Secure the two with cement. Line up the flywheel to connect running position upon the drive shaft. Several light taps on the drive shaft with a rubber hammer will bring the flywheel in line. Now use epoxy cement on the bottom side, opposite the oiled bearing surface. If you let the applied cement set up overnight, the flywheel will remain in its correct position.

Squeaking Noises

A dry or worn bearing will produce squeaks. Check the capstan–flywheel with motor belt removed. If the capstan bearing is excessively worn, replace it. See if the noise is coming from the motor assembly. Remove the belt and apply voltage to the motor. A drop of light oil may clear up the dry motor bearing. If the motor shaft chatters and is excessively worn, replace the entire motor assembly. Figure 6-14 shows several noise sources in a typical player.

Cartridge Problems

In many cases slow-speed problems are caused by a defective cartridge. First, try *several* new cartridges. (It is not uncommon to purchase a new cartridge that is wound too tight or has a bad pinch roller.) Generally, after hours and hours of use a favorite cartridge develops slow and sluggish speeds. The rubber pinch roller may become dry on a plastic bearing; this slows the tape action.

Correct Voltages

The auto tape player was designed to operate from a 12.6V battery supply. When the tape player runs slow in the auto and not on the test bench, check for correct supply voltage. Most autos with the motors running will provide a potential between 12.6 and 14V. Any voltage outside this range should be corrected by an auto electrician.

Reversed polarity may cause the player to run backward. When this happens, you may find several layers of

Fig. 6-13. The flywheel has moved down drive shaft and was scraping against bottom support brace.

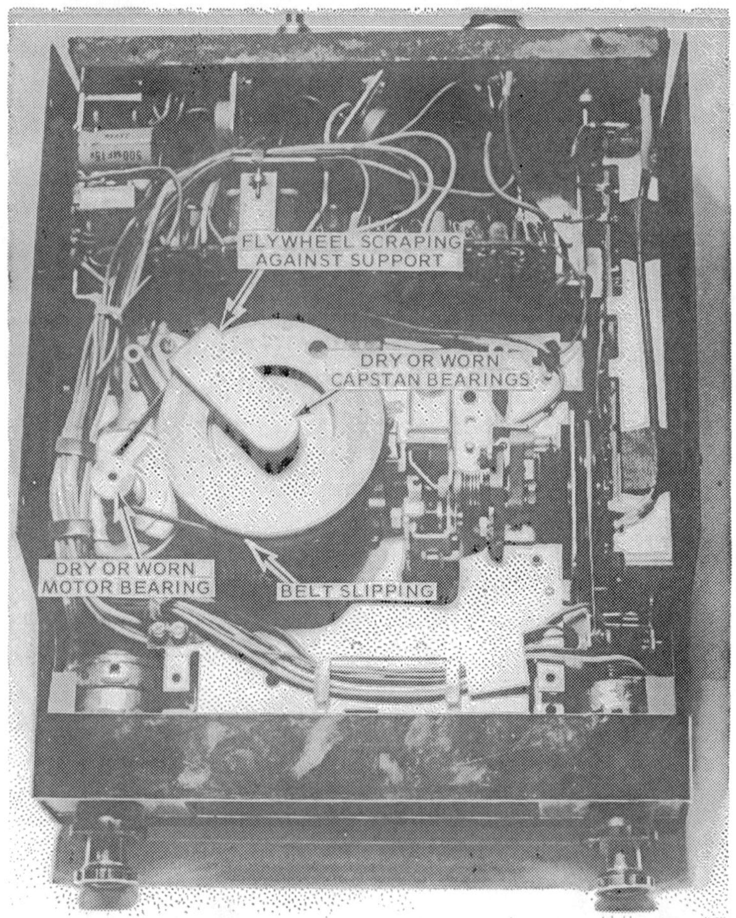

Fig. 6-14. Check for these noisy areas in the stereo 8 players.

tape pulled from the cartridge and wound around the capstan drive assembly. Be careful when connecting power to the player or you may have the same condition. Mark the positive terminal and make sure it connects to the fuse lead. Remember, four-channel stereo units pull more current than the ordinary tape player; these require a bench power supply with nothing less than a 4A current capability. If the bench supply is not delivering sufficient current, the player may run slow.

Troubleshooting the Portable Cassette Tape Player

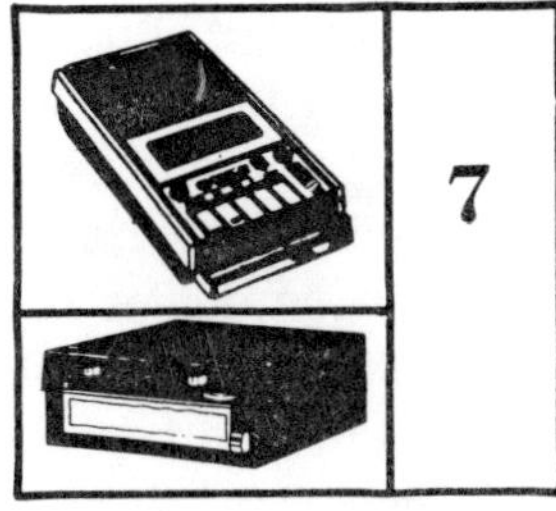
7

You will probably find that servicing cassette players is somewhat easier than it seems. The portable cassette, with its tight-fitting, cramped components, may take a little longer to service than its big-brother counterparts at first; but after servicing a few of them, you'll get the procedures down pat.

PLAYER INOPERATIVE

When the cassette player will not operate, see if any one of the three basic functions—play, fast-forward, or rewind—is functioning. Next, determine if the amplifier has a hiss or a noise with the volume control wide open. If the answer is negative in all cases, the trouble may be that power is not applied to the cassette player. Check to see if it will operate on either battery or ac operation. If not, check the power switch and wiring. (Don't overlook the obvious; the ac cord may be defective; the batteries may be in backwards.)

When the cassette will not play in battery operation, check or substitute new batteries and check correct battery polarity. See if the batteries are seating properly. It's possible the battery holders are warped out of shape. If the batteries have been left in the cassette too long, the contact springs may become corroded. Measure for correct battery voltage at both ends of the battery connecting cables that go to the player.

If the cassette will not operate on ac or from the external power jack, the ac power supply is at fault. First, measure for dc output voltage coming from the power supply. When no dc voltage is measured, check for ac voltage at the silicon diodes. In case of no ac voltage, see if the dc cord or primary winding of power transformer is open. If the ac voltage is found on one end of the silicon diode and no dc voltage is indicated, suspect a leaky or shorted diode. Generally, when both silicon diodes have a direct short and the player has been on for some time, the primary winding of the power transformer will open.

Suspect motor or mechanical problems when a sound can be heard from the speaker with no tape motion. Check for a broken or loose motor drive belt. With your ear next to the player, you should be able to hear the motor run. If not, spin the motor pulley with your fingers and see if it takes off. A dead motor may have worn brushes or an open field coil. Take a voltage check at the motor wire terminals. A frozen capstan–flywheel bearing may prevent the motor from turning, but generally the motor pulley will turn inside the motor belt. Figure 7-1 shows typical trouble areas of a portable cassette player.

CAPSTAN–FLYWHEEL DOES NOT ROTATE

In case the motor rotates but the flywheel does not turn, suspect a loose motor belt or pulley. The pulley is held to the motor shaft with small setscrews. If the flywheel shaft is dry or frozen, suspect a twisted capstan shaft—particularly when the flywheel seems to be binding. This can easily happen when the portable cassette player has been dropped. Check for proper clearance of the flywheel and end bearing. The end bearing may be too tight against the flywheel which will prevent rotation.

NO TAKEUP

The clutch pulley may be slipping when no or poor takeup rotation is noted. You can clean up the clutch pulley with an alcohol-moistened cloth. Adjustment of the clutch arm and shaft may be needed. You can check the takeup torque with a torque cassette or hook the tension gage to the test reel. Check the manufacturer's specifications for proper torque tension in grams. (See Chapter 4 for other pressure adjustments.)

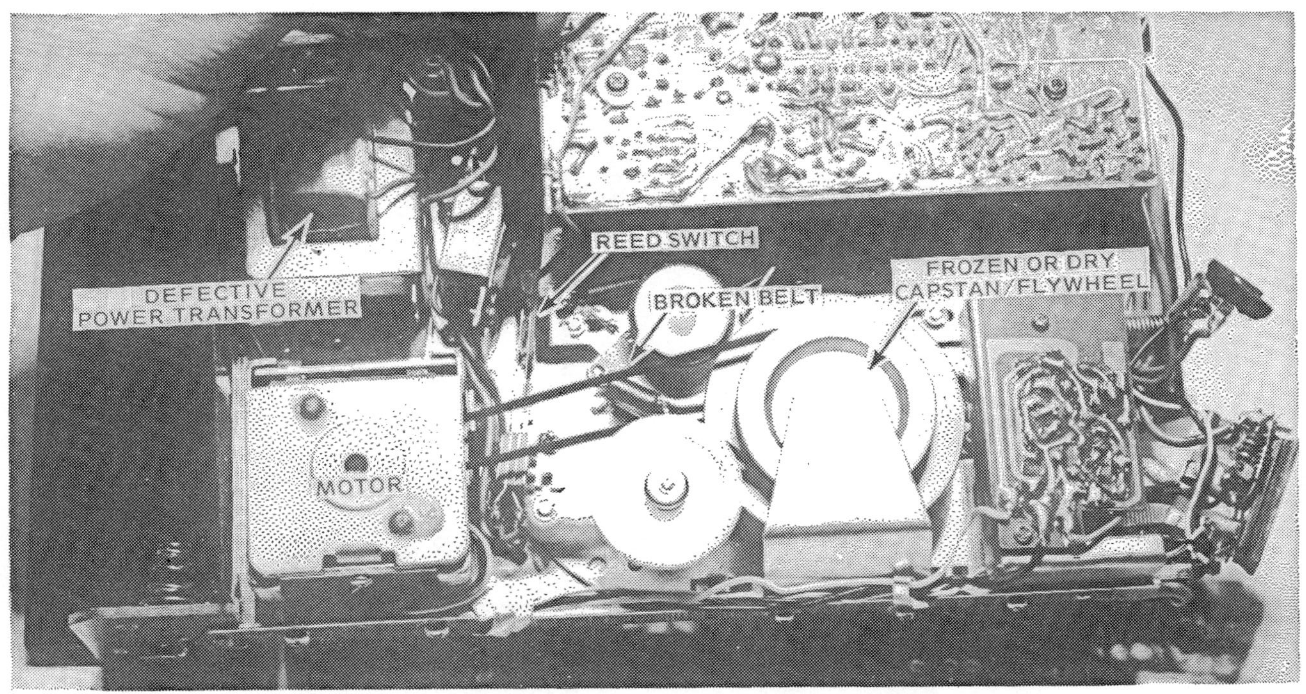

Fig. 7-1. Areas to check when the cassette player is inoperative.

When the pinch or pressure roller lets the tape slip and bunch up, check and adjust the pressure roller spring. On some models, the roller spring can be taken up by slipping it into another hole. If the pinch roller rolls up the cassette tape as a result of a part replacement, you may have to adjust the gap between the pinch roller lever and metal stopper. The bearings of the pinch roller may be dry or the rubber pinch roller may be deformed. A deformed pinch roller may take up the tape unevenly and cause tape slippage. Figure 7-2 shows those areas most commonly causing takeup problems.

NO FAST-FORWARD OR REWIND

In fast-forward or rewind operation the cassette tape should be released from the capstan and pinch roller assembly. The tape, of course, should run at a higher rate of speed in both operations. Check to see if the speed is slow or sluggish. You may find that a good cleanup of belt and rubber drive surfaces cures both problems.

First, clean off the fast-forward pulley with a cloth moistened in alcohol. Clean up the fast-forward idler and rubber drive belt. Sometimes oil will gather on the belt and produce erratic speeds. Check for a worn or stretched drive belt. Once again, try both fast-forward and rewind operations. A drop of oil on the pulley and idler wheel bearings may improve speed conditions.

In models that use a separate motor for fast-forward and rewind operations, check for oily motor belt. When the tape will not rotate in either direction suspect a defective motor. Check the pushbutton assembly for proper seating.

If the brake assembly does not lift clear of both pulleys, erratic or no rotation will be noted. Sometimes one brake shoe may be cockeyed, thus preventing spindle rotation. The brake assembly may be bent out of line and will not slide out properly. Check for a worn slot in the brake assembly. Excessively worn or deformed parts should be replaced with exact part numbers. Often these small components are very inexpensive and should be replaced. Of course, the biggest problem is in stocking and knowing where to obtain the defective parts.

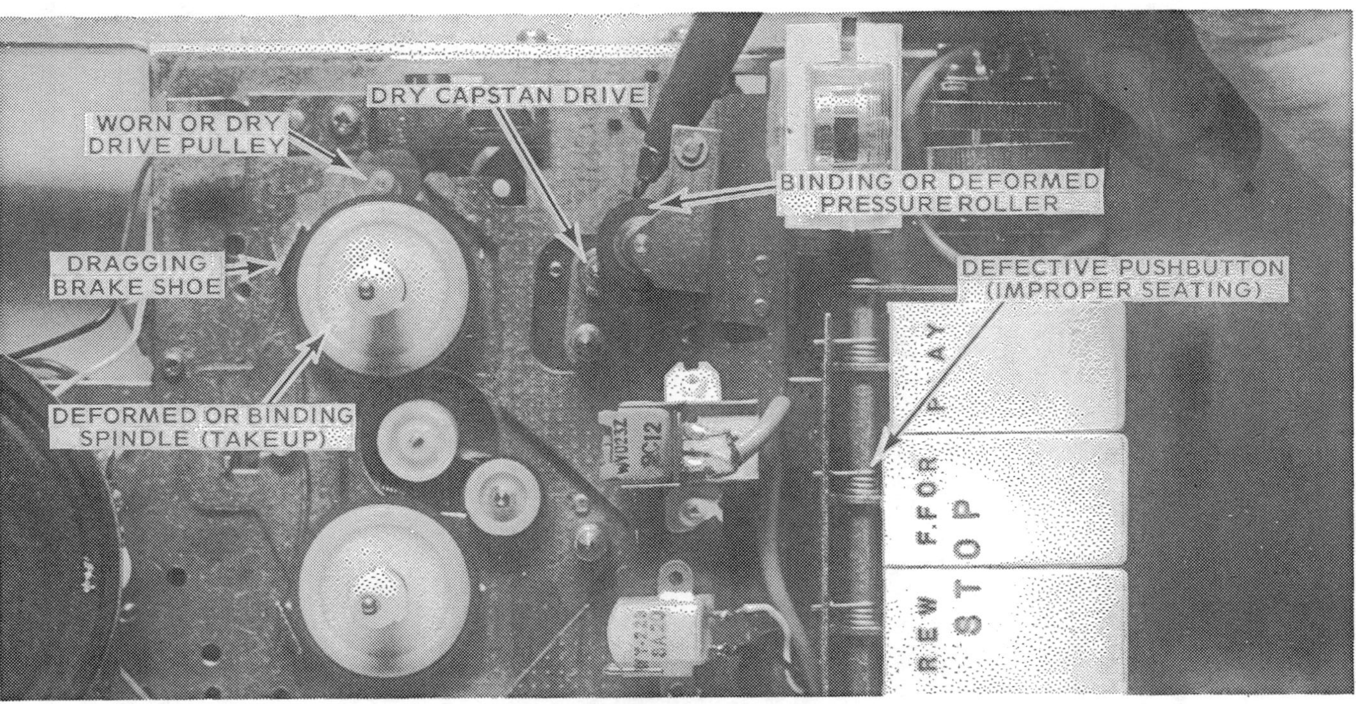

Fig. 7-2. Areas to check for no takeup action.

Figure 7-3 is a photo of a typical portable cassette player; this will help you locate common trouble areas.

EXCESSIVE WOW CONDITIONS

If the customer complains of the recording "not sounding right" or running slow, the trouble could be excessive wow. Such problems are amplified when a malfunctioning machine records and plays back the same tape. Try a new cassette or several to see if the cassette is defective.

After you have determined that the wow problem is in the cassette player, check the most obvious components first. Clean off the rubber drive belt. A worn or oily drive belt will cause all sorts of speed problems. These belts are cheap and should be replaced every year. An oily or dirty capstan isn't the only source of wow problems, as shown by Fig. 7-4.

Clean off the excessive oxide buildup on the pinch or pressure roller. The pinch roller should be cleanedevery time the tape head is wiped off; use a separate cotton swab. Sometimes these rubber rollers will become dry and need lubrication.

Excessive clutch-arm spring pressure produces wow. Check the pressure or pinch roller for insufficient spring tension. A burned-out or dry flywheel bearing is another common culprit. It is best to pull the flywheel—capstan out of its mounting and do a thorough job of cleanup and lubrication. Excessive brake pressure should be checked with a tension gage. You may find the motor drive bearing dry or binding. Don't overlook a defective motor that changes speeds under load. Sometimes a defective tape counter will pull or create drag, too.

PAUSE DOES NOT FUNCTION

The pause lever is not functioning properly when the pause button is pushed and the tape still moves, even if only for a short distance. The pause lever should pull or lift the pinch roller away from the capstan shaft, stopping tape travel instantly. With no power applied, push the pause button in play mode. See if there is a clearance of over 20 mils or so (0.020 in.) between pinch roller and capstan shaft. If the

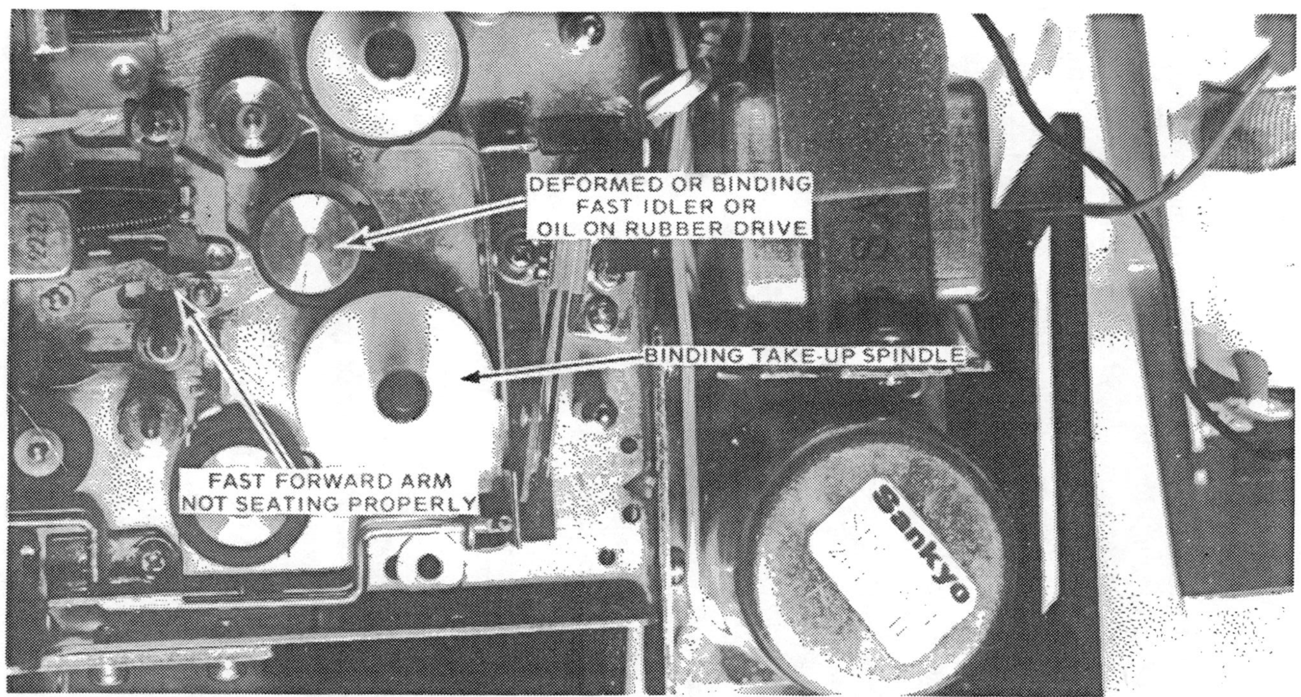

Fig. 7-3. Areas to check for no fast forward or rewind operation.

Fig. 7-4. Areas that will cause wow conditions.

clearance is less, bend the pause operating lever. Now push the pause button again slowly and confirm that the pinch roller and takeup reel stop simultaneously. If not, check the pause button and guide lever for alignment.

SUPPLY SPINDLE DOES NOT ROTATE

In case the supply spindle does not rotate or runs too slow, check to see if the rewind operation is normal. The supply reel spindle is on the left; it operates in the play, record, and rewind operations (Fig. 7-5). If the supply reel does not operate properly in either direction, see if the supply spindle will turn by hand. The supply spindle bearing may be dry or worn. Check to see if the plastic supply spindle is deformed. With the cassette removed, see if play and fast-forward operations appear normal.

The fast-forward/rewind roller pressure may be insufficient, thus preventing normal supply spindle rotation. Adjust the pressure spring or install a new one. Check the fast-forward/rewind rubber wheel for oil or deformed areas. In rewind operation the rewind "tire" may be soiled or

Fig. 7-5. The supply spindle unwinds tape and rewinds tape in rewind operation.

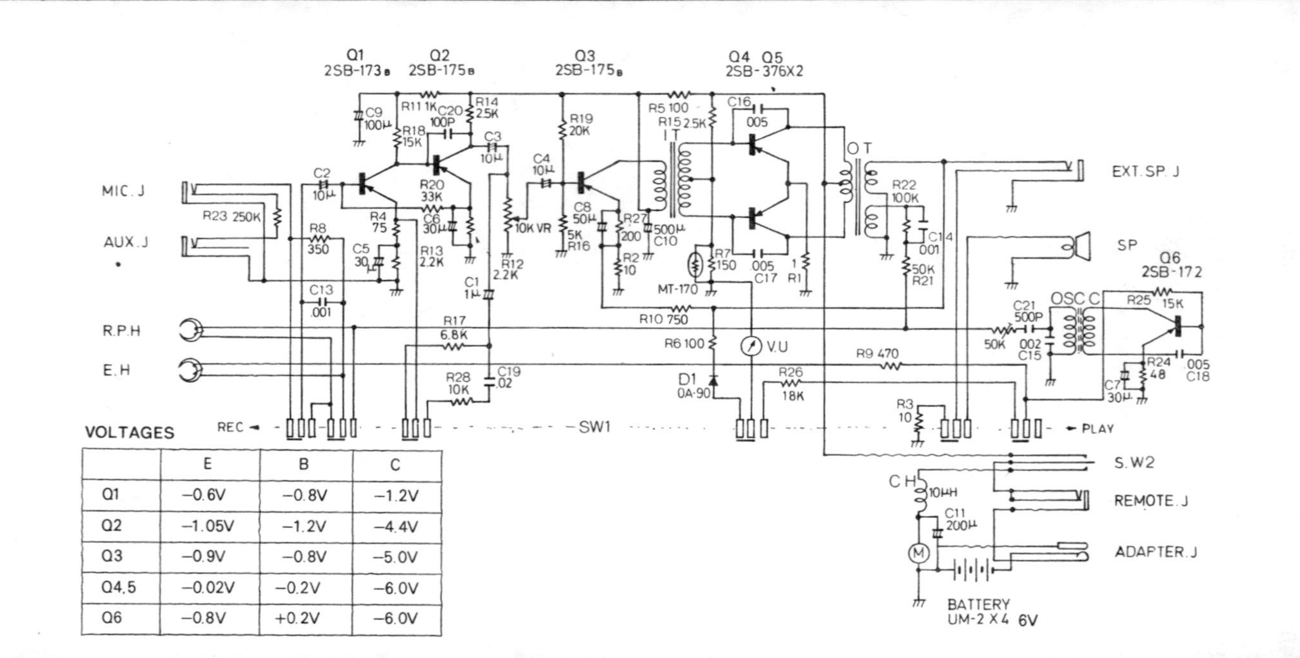

	E	B	C
Q1	−0.6V	−0.8V	−1.2V
Q2	−1.05V	−1.2V	−4.4V
Q3	−0.9V	−0.8V	−5.0V
Q4,5	−0.02V	−0.2V	−6.0V
Q6	−0.8V	+0.2V	−6.0V

Fig. 7-6. Small portable cassette motor circuit. (Courtesy Automatic Radio.)

deformed. Also, see if the brake assembly is releasing from the supply spindle. A locked or dragging brake will produce very little supply reel rotation. In case the supply reel spindle rotation is not smooth, check the motor pulley, flywheel, and rollers for oil spots.

CASSETTE MOTOR CIRCUITS

The motor circuits found in the portable cassette players are typically quite simple. One is shown in the cassette player schematic of Fig. 7-6. Since the motor operates from batteries or an external power supply, only a small amount of current is used. The internal batteries are in the circuit until the external power supply is plugged into the adapter jack.

In some portable cassette players, the power supply is built-in. Normally, the supply is a simple affair similar to the circuit of Fig. 7-7. Components T1, CR2, CR3, and C4 form the low-voltage power supply circuit. Switch SW1 is a simple reed switch. Switch SW2 selects either battery or ac operation.

Motor Does Not Rotate

If the motor does not rotate in any pushbutton operation suspect a defective motor. Remove the drive belt and turn

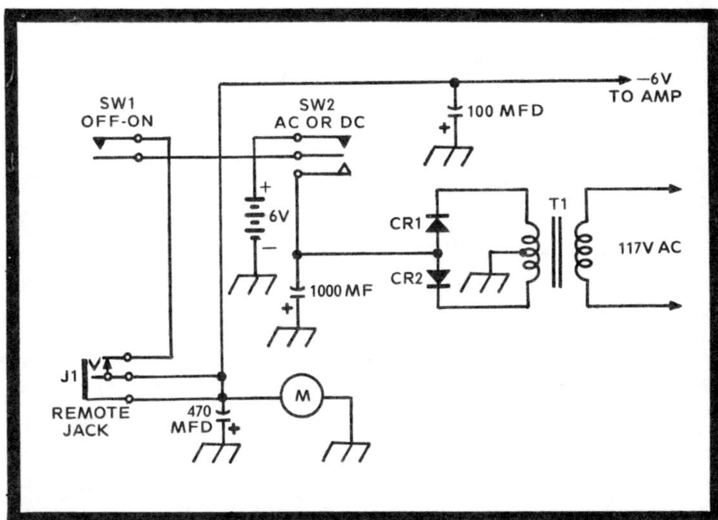

Fig. 7-7. Portable cassette operates on battery or built-in ac supply.

the motor pulley by hand. A dry or frozen motor bearing may be the problem. When manually rotating the dc motor you will notice a pull and quick forward motion. Inside the motor is a permanent magnet that will appear sluggish when you turn the armature metal area near it. This is normal. See if the motor will rotate without a load. Check for correct motor voltage at the motor wire terminals.

Noisy Cassette Motor

There are two types of noise that may be troublesome: *mechanical* and *electrical*. When an electrically produced noise appears, make an effort to determine the source. Electrical noises can be caused by external sources (as with interference from ignitions, motors, arcing motor brushes, etc.) and internal sources. Internally generated noises are caused by heavily magnetized heads, intermittent transistors, poor electrical connections, and any of a dozen or so other anomalies. When you've determined the noise is internal and electrical start your trackdown effort by demagnetizing the tape heads.

Sometimes you'll have what amounts to a dual problem—electrical *and* mechanical noise in the same tape player. In this case, the first suspect area will be the motor. As mentioned earlier, arcing brushes and improper brush-to-armature contact can cause severe audio interference. And usually, when a motor's commutator is worn or when the brushes appear to be wearing unevenly, a scraping or rasping sound can be detected at the motor itself.

Remove the brushes and examine the ends that contact the commutator. Are they worn evenly, following the curvature of the armature? Are they flattened on one edge? Are they chipped? If examination reveals a problem here, either replace the old brushes or reseat them. It's also a good idea to examine the commutator for evidence of flattening, arcing, carbon buildup, and the like. If you're reseating the old brushes, it's usually wise to sand the commutator a bit with a very find grade of sandpaper.

Be very careful not to inflict your own damage to the brushes. They're fragile and tiny—and they break easily.

If the motor noise is not exceptionally loud, you can connect a 200 μF or larger filter capacitor across the motor terminals and eliminate the noisy problem. You may try this method if the motor replacement is going to be costly.

Several small retaining screws are typically used to hold the motor in place. The motor frame is often slotted (Fig. 7-8) so the motor can be turned to allow limited belt adjustment. Loosen the retaining screws. Switch the unit to playback mode. Try rotating the motor to adjust its position until the noise is minimized. Then retighten the two retaining screws.

NO SOUND EXCEPT NOISE

Determine the source of the noise first. Is it coming from the amplifier? Turn the volume control to minimum and maximum with a blank tape inserted, and listen for the noise. If the noise is only present when the volume is wide open, suspect a magnetized tape head or noisy preamp stage. Short the tape head terminals. If the noise is still present, the problem is very likely between tape head and first audio section.

Take a quick look at the wire connections of the tape head for loose connections. Inject an audio signal on the ungrounded terminal (inner conductor of shielded wire) of the tape head. A pencil-type noise generator (Fig. 7-9) is ideal here. Proceed from tape head to the base of the preamp

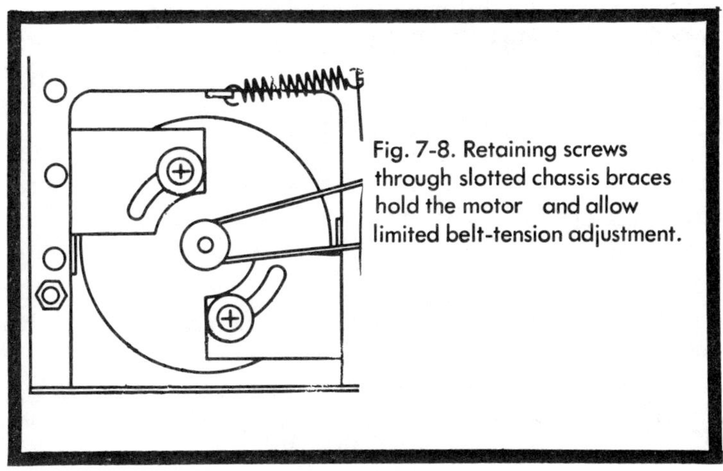

Fig. 7-8. Retaining screws through slotted chassis braces hold the motor and allow limited belt-tension adjustment.

Fig. 7-9. You may use a pencil type noise generator to test the amplifier at the tape head.

transistor. A leaky or shorted electrolytic coupling capacitor from tape head to the base of the preamp stage may produce either very weak sound or no sound at all.

If the preamp stages are direct-coupled, go to the collector of last preamp transistor. If the generator signal wasn't heard when it was applied to the head terminals, but *is* heard at the speaker when applied to the collector of the final preamp stage, you've located the trouble: a defective preamp stage. Simple voltage checks should point out the defective transistor or component in short order.

Be careful in placing the probe of an audio generator so as not to short out (and thus destroy) a transistor or integrated circuit. A short circuit between base and collector can destroy a transistor.

If the sound comes on when the collector of a transistor is touched, but does not come on with the probe touching the base, the problem is automatically isolated to that transistor.

When the problem is intermittent, use "cold" spray on the transistor and see if it acts up again.

NO PLAYBACK

Before tearing into the cassette player when the symptom is *no playback,* see if the player will record. If the tape player records and will not play back, the trouble lies in the speaker or headphone jack assembly.

What you must do now is determine whether the problem is mechanical or electrical. If the tape transfers properly in playback from the supply reel to the takeup reel, start looking for an electrical problem. Turn the volume control all the way up with a known-good recorded cassette in place. Do you hear any sound—hum, rushing noise, etc.? If there is *any* sound at all, it indicates the power supply for the transistor circuits is functioning, and that the final amplifier, at least, is delivering a signal to the speaker. If no sound is heard—but the tape does move—it's a sign of a malfunction in the amplifier section, speaker, or power leads connected to the amplifier.

First, take the unit out of its case. You can refer to the schematic in Fig. 7-10, which is fairly typical of a portable cassette player. Note that switch SW2 (far left, adjacent to head labeled R&P) is in the playback position. Make sure that one of the two head terminals is grounded, either directly or through the contacts of a switch. Once you've determined that one side of the head is indeed grounded, touch the other terminal (volume at maximum) with a finger or a long metallic probe. If you hear hum at the speaker when you do this, you've isolated the problem to the head. Either the signal lead has become disconnected from the head, or the head is defective.

Broken leads at the head terminals are very common. With a cassette player, the head is moved against the tape and away from the tape by a mechanical switching arrangement (to allow fast-forward and rewind operations without the head coming into contact with the tape). This continual movement frequently causes one of the terminal leads to break at the point of connection. Since this occurs far more frequently than any other head anomaly, it's very important to check for this possibility.

If you touch the signal lead at the head and don't hear a corresponding hum in the speaker, you'll have to do some

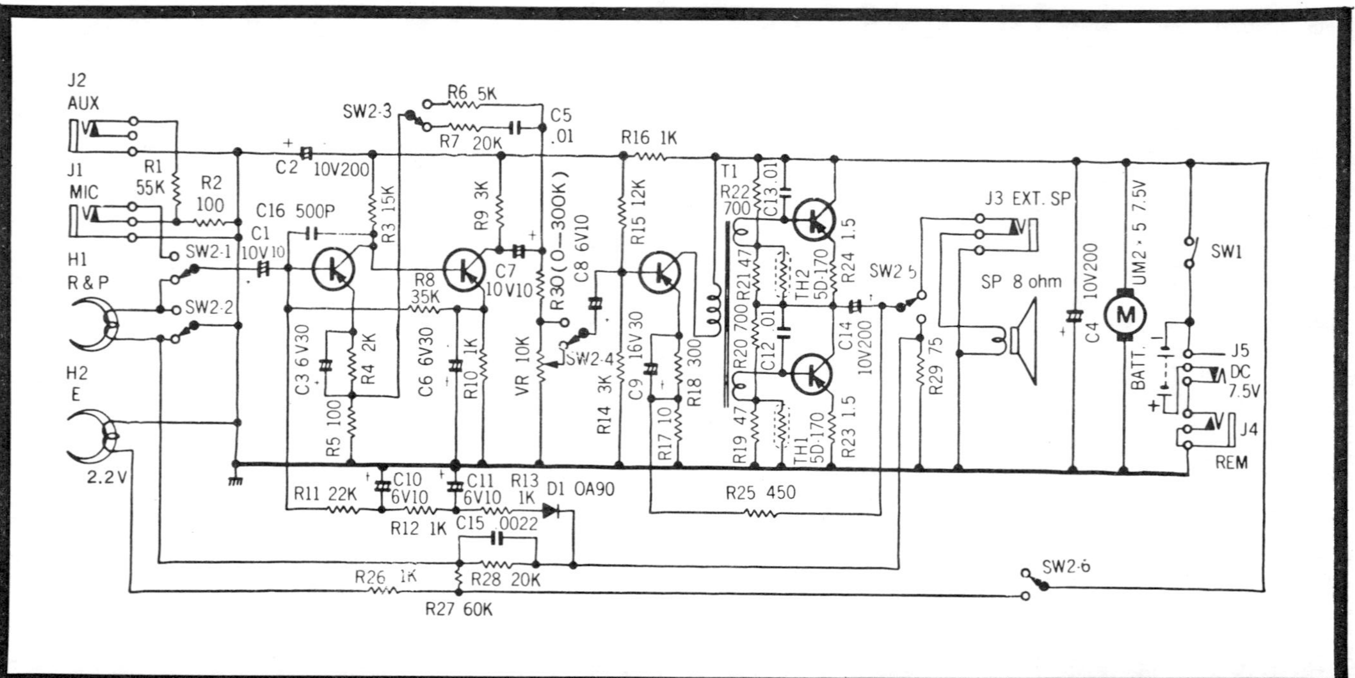

Fig. 7-10. Typical portable cassette-player schematic.

troubleshooting. The halve-and-try technique usually works out to be a timesaver, so start with that approach after you determine that supply voltage for the transistors is available. Start at the base of the driver transistor (in the schematic, this is the transistor shown just to the left of the push-pull pair). Apply a signal here from an audio generator (any audio oscillator will work for this series of tests, since all you're trying to do is get a signal from the generator to the speaker). Hear the signal at the speaker?

If so, the trouble lies somewhere between the head and the transistor you've just applied signal to. If not, the trouble lies between the transistor you're touching and the speaker.

For the moment, let's assume that you did not hear the signal at the speaker when you applied signal to the base of the driver transistor. This tells you there is a problem between that stage and the speaker. Move the probe to the collector of the driver stage. If you get signal now, replace the driver transistor and the set is fixed. If you don't hear the signal, you should start checking the wiring between the driver and the speaker. The output transformer could be bad, but it's rather uncommon. And if one of the two output transistors is bad, you should still be able to hear speaker audio—it will be distorted, of course, but *there* nonetheless. The most likely trouble will be a bad connection in the switch (SW2-5), plug J3 (*very* likely), or the speaker itself. The electrolytic (C14) could be leaky, but odds are against it.

Now let us assume that you *did* hear a speaker signal when you applied the generator output to the driver transistor's base, which indicates the trouble lies between that point and the head. Use the halve-and-try technique again, moving the oscillator probe to the transistor immediately ahead of the driver. You can see the philosophy of the approach by now. Simply move the audio probe toward the head, a bit at a time, until the signal disappears. When it does, you've isolated the malfunctioning stage.

Don't overlook a poor contact in the short-circuited muting switch. Check the speaker out with the ohmmeter. When the ohmmeter is on the low ohm scale you can hear a click in the small speaker when touching speaker voice coil. You can quickly check the earphone jack and speaker by placing the ohmmeter lead to collector or output terminal and chassis ground. When you heard a loud clicking noise the speaker jack is making good contact. If not, suspect a poor contact of the short-circuit earphone jack. When making these tests with an ohmmeter make sure the power is turned off.

WEAK OR LOW SOUND

First, check to see if the weak reception occurs in battery or ac operation (or both). If only in battery operation, suspect weak batteries or corroded terminals. When weak condition is noted in both operations, check for a dirty tape head. Try another cassette before removing the chassis from the cabinet. After the tape head is cleaned and another cassette is inserted, see if the player has weak reception in both record and playback operations.

Most *weak* conditions are combined with excessive distortion. You may suspect a defective tape head or amplifier circuit. A worn record–play head will produce weak reception. Improper bias current and frequency can also cause weak conditions.

DISTORTION

Distortion is quite common in cassette players. Check to see if the distortion is found in both battery and ac operation. Insufficient supply voltage (weak batteries) will produce distortion. After cleaning up the head, see if the distortion is noted in playback *and* recording operations. Try a new prerecorded cassette in playback operation. When distortion occurs in both playback and recording operation, suspect a defective or dirty tape head or amplifier problems.

When excessive distortion is found in only the recording mode suspect improper or no ac bias voltage to the tape head. Check the bias signal at the head terminals with the oscilloscope. Place the scope's *direct* probe on the ungrounded side of tape head. You should see a sine wave at this test point. If not, check the oscillator bias circuit for no or low signal.

If distortion is found in both playback and record operations, signal tracing of the amplifier circuit is in order.

Improper adjustment of the bias resistor of audio output amplifiers will produce distortion. A poor wiper contact on a bias resistor has caused many a distorted amplifier section.

Some unusual distortion problems apparently attributed to a fault in the amplifier section are not even caused by any component in that section. For example, a new Panasonic cassette player (RQ-226S) had distortion in both record and playback operation. Distortion was apparent with both battery and ac operation. The distortion was traced to the audio output stages.

A visual indication showed no burned bias resistors. Then I suspected leaky output transistors. Voltage measurements on the output transistors were way off, so I decided to remove the output transistors. When removing the heatsink, I found the transistor leads were crossed and pressure from the heatsink was biting into the insulated transistor leads (Fig. 7-11).

After removing the heatsink, the cassette player was switched into operation with good volume and no distortion. The cassette player was only 60 days old, so my assumption

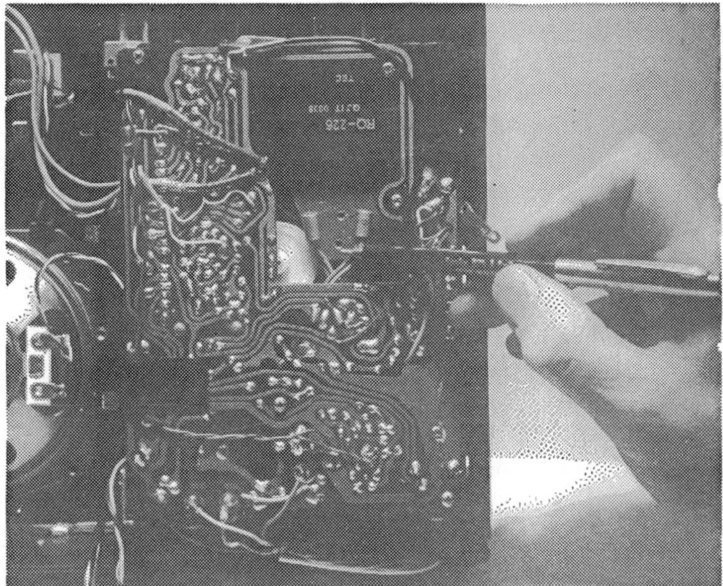

Fig. 7-11. A unusual type of distortion was found with the heatsink cutting into the output leads of the push-pull transistors.

was that carrying and jarring caused the heatsink to short out the output transistor leads.

Intermittent distortion problems are difficult to locate in any amplifier section. It may take several hours and days to locate the most difficult one. But by employing time-tested isolation methods and using your scope effectively you can lick any distortion problem. Isolate the distortion to a certain section. Inject sine and square wave signals from the audio generator and clip the scope leads to the various components until the intermittent problem acts up. Check off each suspected component until the culprit is located.

WILL NOT RECORD

Before removing the chassis see if the cassette player operates normally in playback. If it does, we can rule out a defective amplifier and tape head. In most cases one tape head and the same amplifier serve for both play and record operations. Talk into the microphone and watch the VU meter for voice indication while recording. When the VU meter fluctuates while talking into the microphone you know the microphone and amplifier are okay. The problem may be the record switch, a defective oscillator, or record tape head.

But there's yet another potential problem area. Examine the circuits presented in Figs. 7-10 and 7-12. Note the switch contacts adjacent to the recording head (R&P). In both players the head-sharing philosophy is the same—in one position, the lower terminal of the head is grounded; and in the other position, the upper head terminal is grounded. Both players are shown in the play mode. Note the long lead connected to the lower terminal of the head in both players. If this breaks—and it will on occasion—the player will work perfectly on playback, since that terminal is grounded in this mode anyway. But in the record mode, when the switch is moved to the up position, the ground is moved to the upper head terminal and the lower terminal becomes the signal point. Since the signal to the record head must pass through that long broken lead, the unit will be inoperative until the break is repaired.

Be sure and check the record–play switch for poor contacts and connections. Simply run the small tip of your

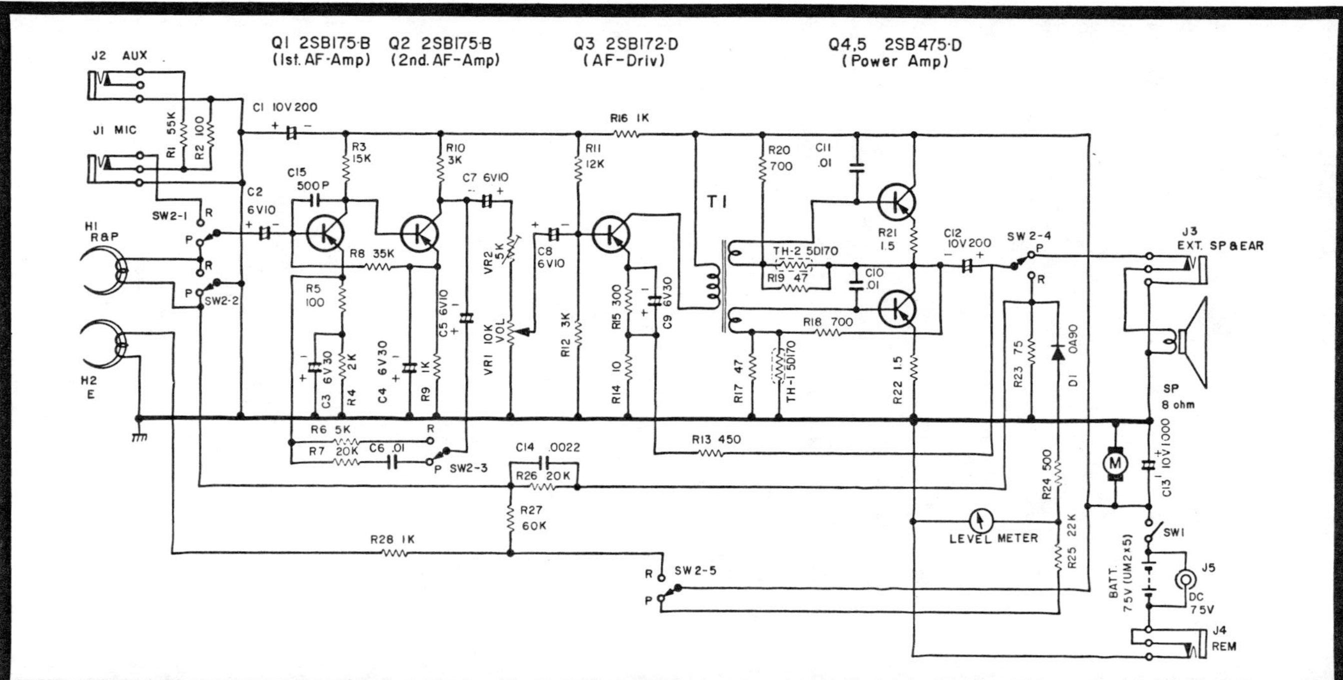

Fig. 7-12. Portable cassette recorder—player with two-stage preamps. The third transistor drives the push-pull output pair.

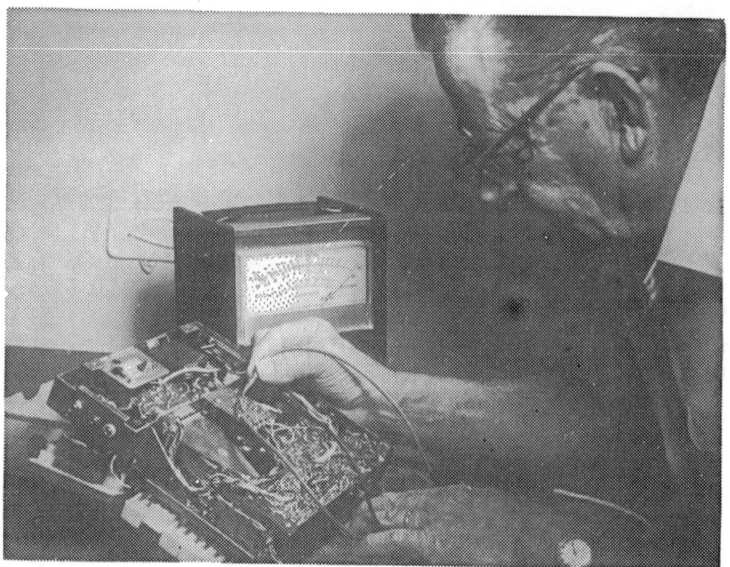

Fig. 7-13. The oscillator bias transistor may be checked in the circuit with an in-circuit transistor tester.

soldering iron over the switch contacts on the PC board. You may find a break or poor solder joint at these connections. Spray contact cleaner down inside the record–play switch as you push the switch back and forth. A good switch cleanup may solve both play and record problems.

A defective component in the oscillator bias circuit can prevent good recordings. If no bias signal is present at the tape head (check with scope), look for a defective component in the oscillator section. Take voltage readings on the oscillator transistor, too.

Figure 7-13 shows a quick and easy way of checking the oscillator transistor—it's an in-circuit beta transistor tester, and a particularly handy gadget to have in the shop if you do much circuit-sleuthing. Another possibility is the oscillator coil. Make continuity checks of the oscillator coil windings. The oscillator coil and transistor are a common cause of problems in any bias section. (Improper bias frequency can cause poor or distorted recordings.) Replace the tank capacitor across the primary winding of oscillator coil.

If the cassette player operates in playback and does not have a VU or recording meter, suspect a defective microphone cable or jack. The mike itself could be bad, of course, but by far the largest percentage of input-circuit problems can be traced to mechanical defects. Inject an audio signal in the microphone jack and see if it will record. Sometimes the microphone jack will short to the chassis or open because of poor contact. Look for evidence of a broken wire on the mike jack.

The suspect microphone can be checked with the ohmmeter. Simply touch the microphone male jack tip and ground area with a VOM on the R × 100 scale. You should hear a click in the microphone like that of a PM speaker, and see an indication on the meter. If not, the mike coil winding or cable is open. If the click is not apparent and the meter deflects fully, the cable is shorted. Generally, the microphone cable will break where it enters the body of the mike or at the male plug end (Fig. 7-14).

If the complaint is that a certain cassette will not record, see if the small notch at rear of cassette is knocked out. All prerecorded cassettes or test cassettes come with slotted areas on each end and backside of the cassette. This prevents

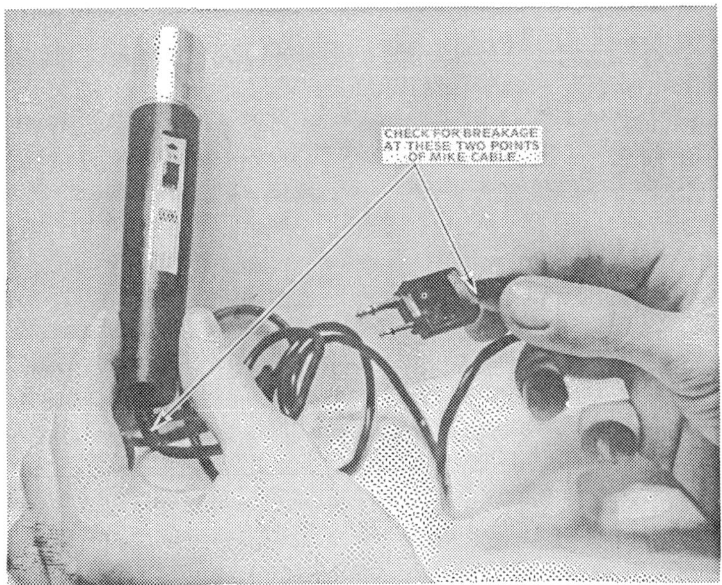

Fig. 7-14. A microphone cable usually breaks at the body of the case or where the cable terminates at the male plug.

erasing of the prerecorded cassette. Generally, the complaint is cassette player won't record on a certain cassette and the owner forgets the small notch has been knocked out. Just place a thin strip of vinyl or cellophane tape across the knockout area to remedy this situation.

ERASE PROBLEMS

When the customer complains the recording is garbled or more than one recording is heard at one time, the erase head is not functioning. Take a brand new cassette and make a short musical recording. You can place the microphone near a radio for this recording. Rewind the cassette and see if the recording is normal. Then rewind the tape again to the beginning and speak into the microphone in the record position. Rewind the tape and see if both voice and music can be heard in the recording. If the music portion is heard, the erase head or erase oscillator circuit is not working properly.

Check to see if the erase head is clean and fits snugly against the tape. Determine if the suspect tape head is magnetic, dc operated, or excited with an oscillator stage. A small screwdriver blade placed near the magnetic erase head will pull the blade to it.

An erase head excited by a dc voltage or oscillator stage will have a connecting cable wire to the erase head terminals. In record operation see if a dc voltage can be measured across the erase head terminals. If not, the dc voltage is missing or the erase head requires an oscillator signal to operate. Now, check for a sine wave (Fig. 7-15) on the erase head terminals with an oscilloscope. Remove one lead from the erase head and see if the winding is open.

When erase head problems become intermittent servicing the cassette player will take a few hours longer. For example, in an AM–FM cassette player, excessive crosstalk was noted. After checking with a couple of known-good cassettes the chassis was dismantled and partly removed from the cabinet. This particular intermittent erase problem took several hours to locate the defective part.

We measured 1.5V ac upon the erase head with a VTVM. The erase head alternating voltage in cassette players of this type should measure between 13 and 15 volts. When the scope leads were coupled to the erase head the bias voltage increased to a normal 14.5 volts ac. Then the cassette player would operate all day long. After sitting overnight the player was defective again.

Each time the erase head terminals were accidentally shorted the bias erase voltage would return to normal. Of course, the trouble had to be an intermittent component in the oscillator bias section. The scope was clipped to the erase head terminals and a VTVM tied to the oscillator voltage supply source (−7V). The next morning the sine wave was weak and the −7V supply voltage was okay. Voltages upon the oscillator transistor were near normal when in the intermittent condition. Now we knew the supply voltage (−7V) was not the problem.

The oscillator transistor was removed and tested. A new one was installed and the unit played perfectly. Yet, the next morning the player was intermittent. The oscillator coil continuity was good. All resistor measurements were accurate. After the third day, replacement of the 100 μF electrolytic capacitor solved the intermittent crosstalk and

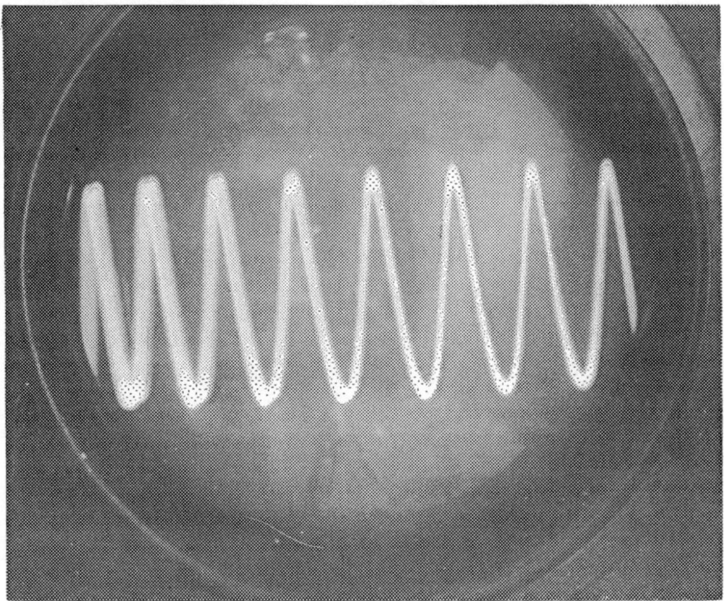

Fig. 7-15. If the oscillator circuit is performing, you should see a sine wave at the erase head terminals.

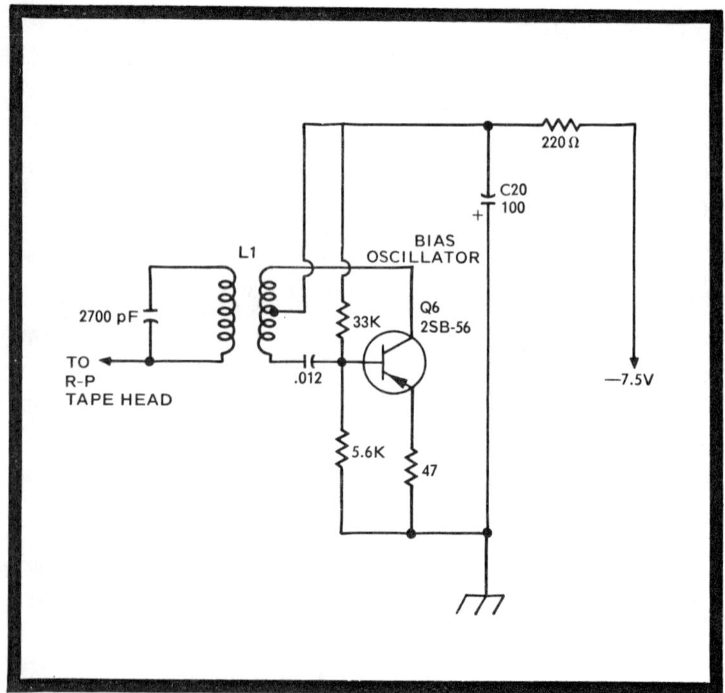

Fig. 7-16. Capacitor C20 solved the intermittent erase problem.

distorted recordings. In the circuit shown in Fig. 7-16, the problem capacitor was C20.

Although this erase problem was difficult to locate, most erase and bias oscillator conditions are quite simple to service. You can locate most bias problems with the scope and VTVM.

REMOTE PROBLEMS

If the player will operate without a remote microphone and not with the remote plugged in, suspect a defective cable or jack. The remote power switch may be defective or have dirty contacts. Slide the remote switch to the on position and check continuity with ohmmeter. Dismantle the microphone assembly and measure resistance across the switch terminals. You will find most cable breaks will occur where the cable enters the microphone case or at the male plug. Don't forget to check for a broken or bent jack inside the player. You can locate most remote problems with the VOM.

EXCESSIVE HUM

Most hum conditions are found in the front end of the amplifier or power supply of the cassette player. First, isolate the hum problem to play or record operation. Turn the volume control all the way down in play position. Notice if the volume control is excessively noisy when turned up or down. A worn or open spot within the volume control can cause an erratic hum.

If the hum disappears with the volume control turned down, this indicates hum pickup in the tape head or preamp stages. A defective tape head or improper terminal connection will produce hum. Flex the microphone cable and check for a broken or poor male plug connection in the record position. Check for poor microphone terminal connections at the microphone jack.

With the volume control turned down, if you notice excessive hum in play position, suspect an electrolytic filter or decoupling capacitors. Large filter capacitors paralleled across the suspect ones will eliminate power supply hum problems. Be sure and observe correct filter capacitor polarity. Do not bridge them across when the power is on. A dried-up decoupling capacitor may produce hum and oscillations.

Excessive hum may be found in the power output transistor stages. Remove the suspect transistors and check for leakage. If in doubt, replace them—it is the quickest and the best method to locate hum in these stages. Generally, output transistor hum is accompanied with some type of distortion. One transistor in a push-pull or complementary-symmetry output stage may be open and produce the same results.

CASSETTE EJECTION PROBLEMS

A few cassette tape recorders have a feature that automatically ejects the cassette at the end of the tape or when the power is turned off. Also, the feature is found in cartridge tape players. This special circuit senses the end of the cassette or when the power is turned off and ejects the cassette. Also, in some models, when the cassette is ejected at the end of a tape cassette, the player is turned off.

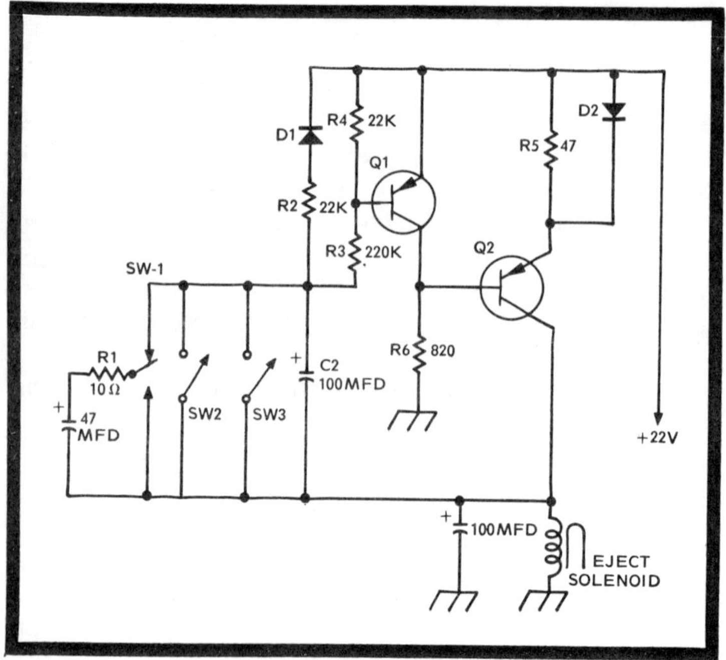

Fig. 7-17. Simple cassette ejection circuit.

Figure 7-17 shows a simple circuit found in some automatic cassette eject operations. The solenoid ejects the cassette when energized by the conduction of Q2. When the tape is moving, the rotating wiper of SW1 alternately closes the contacts of the switch. SW1 is part of the tape supply spindle. When the tape motion stops, switch SW1 can be open or touching either contact. As long as the tape is moving, the cassette will not be ejected. With the tape moving, Q1 conducts because increased base current is available. The base current of Q1 charges capacitor C2. Capacitor C2 never reaches its full charge as SW1 alternately connects C1 across C2 and discharges it. With the tape in motion a continuous base bias is found in Q1. If the tape stops, SW1 no longer turns, letting C2 assume a full charge. When C2 is fully charged, Q1 has no base current and is cut off. With Q1 not conducting, Q2 conducts and current flows through the eject solenoid. The energized solenoid ejects the cassette. When the cassette is ejected, SW2 closes and discharges C2. This prevents the cassette from being ejected if another one is immediately inserted. Diode D1 forms an alternate discharge path for capacitor C2.

When the power is switched off, the cassette is ejected because the emitter voltage of Q1 lowers much faster than the base voltage due to the charge of C2. Within a few seconds Q13 switches off after the power is turned off and Q2 draws current, thus energizing the eject solenoid. In this same circuit a pause switch may be found in some models, which allows the operator to stop the tape in motion. The cassette will not automatically eject when the power is turned off as long as the pause button switch is closed.

The biggest problem with most automatic cassette ejection systems is the cassette keeps ejecting and will not play. Save yourself some service time and clean the rotating commutator and wiper contacts. This will solve most quick ejection problems. See if the eject lever arm is bent out of line and will not reset. Take voltage measurements on transistors. You may remove one transistor from the circuit and test with the transistor tester. The remaining transistor can be checked in the circuit with an in-circuit beta tester.

OTHER CASSETTE PLAYER PROBLEMS

You should always demagnetize the tape heads after the player has been serviced. Select either a demagnetizing unit that will go through the cassette opening or use a demagnetizing cassette cartridge. Not only will a good demagnetizing job eliminate excessive noise, it will improve the frequency response of the tape head.

Every ten hours of operation the cassette tape head should be cleaned to insure longer life and improve the quality and clarity of sound production. You can easily insert a cassette head cleaner for this operation. Audio Magnetics Corporation makes a cassette head cleaner—with nonabrasive polishing action—of polyester fibers. It is an endless cassette cartridge. The polishing material has a porous appearance which will absorb or collect oxide dust. This cassette tape looks like compressed fiber glass material.

When excess tape has unwound from cassette and is caught between capstan and pinch roller be careful not to tear or kink the tape. After the tape has been pulled from player, rewind it. Place a pencil inside cassette spindle and turn up excess tape. If the tape is broken use a tape splicer and put it back together.

Sometimes those tiny Phillips screws are difficult to remove. It doesn't matter what condition the Phillips screwdriver blade is in—you just can't remove them. Try applying excess heat from the soldering iron to loosen screw cement. If it still won't budge, slot the head with a hacksaw blade. You can now remove the most difficult Phillip screw with an ordinary screwdriver. Generally, these small screws are made of very soft metal and can be slotted for easy removal.

A Sanyo automatic-reversing cassette player (auto mounting type) was brought into the shop with an interesting problem: it wouldn't stay in the right-to-left mode, although it operated flawlessly when the tape was being transferred from left to right.

It took a little time to find the source of the trouble, as the customer didn't volunteer some vital information (which I'll get to in a moment). The Sanyo reversing system works like this: When tape motion stops while a cassette is in place, the reversing solenoid is triggered automatically at the end of a timed period. This machine has no automatic eject feature at all, and simply reverses the tape at the end of travel or when any problem arises where tape motion is stopped.

The problem in this machine was that tape motion was being stopped periodically in the right-to-left mode because one of the reel clutches was binding. The spindle itself was slightly bent—not much, but enough to create the problem. The eject lever had been drastically bent, which indicated that the customer had pushed the eject button with excessive pressure.

When the customer came in, and was asked about the bent ejection mechanism, he stated that a cassette had fouled during play, and the machine kept winding up the tape. The motion would continually reverse, winding the spoiled tape first one way, then the other. He had tried to eject the tape, but the button didn't work, regardless of how much pressure he applied to it, so he panicked becaused he thought the player was going to be damaged irreparably. He looked for a tool to pull the cassette out, and the only item he had was a huge pair of hedge clippers. He poked the sharp-pointed scissor ends of the hedge clippers into the knockout slots of the cassette that were exposed, and forcefully pulled the cassette free. When he pulled all the tape debris out of the receptacle, there was no visual indication that anything had taken place (although the cassette itself, of course, was damaged beyond repair).

The problem with "no play" in the right-to-left mode was as a direct result of his use of force—when he pulled the cassette, the pressure bent the left spindle, causing it to rub against the clutch and brake pad. A rebending to true the spindle up visually paid off, and it remedied the problem.

The moral of this case history is this: Always be sure to ask your customer as many questions about his problem as you can dream up. Chances are he'll be glad to talk about it, and it can save you hours when you're troubleshooting the innards of an unfamiliar machine.

CASSETTE PLAYER TROUBLESHOOTING CHART

TROUBLE	CAUSE
Inoperative	Defective power switch
	Defective batteries
	Defective power supply
	Broken or stretched flywheel belt
	Defective motor
	Frozen capstan flywheel
Takeup hub does not wind tape	Slipping clutch pulley
	Dirty pinch roller
	Slipping pinch roller
	Clutch arm out of adjustment
No fast-forward	Rubber belt slipping or worn
	Idler of fast-forward pulley slipping
	Deformed fast-forward spindle
	Defective fast-forward motor
	Defective cassette

TROUBLE	CAUSE
No Rewind	Rubber belt of rewind slipping or stretched belt Rewind idler duty Deformed supply spindle Defective brake
Erratic speeds	Rubber belt slipping on flywheel Insufficient pressure of pinch roller on tape Dry capstan bearing Defective motor Defective motor pulley
Excessive wow and flutter	Flywheel bearing dry or worn Stretched rubber belt Deformed pinch roller Oily capstan, pinch roller, and belt Insufficient pressure of pinch roller Too much brake pressure Defective tape counter belt Defective clutch assembly Sticky pinch roller
Pause does not function	Pinch roller sticks Sticky clutch Defective pause spirng and lever Deformed slip roller lever
No sound	Open or shorted head Dirty oxide pocket tape head Defective amplifier component Defective power output transistor Poor muting switch contacts
Weak or distorted output	Improper bias current Improper bias frequency Defective amplifier components Leaky or shorted power output transistors

TROUBLE	CAUSE
	Burned or open emitter and base resistors Dirty tape head Worn or defective tape head
No record	Defective head changing switch Open or shorted tape head Open oscillator bias coil Defective components in oscillator circuitry Defective microphone Defective microphone input jack Defective mike cable Dirty head
Distorted record	Defective microphone Improper bias current Bad oscillator bias components
No erase	Open erase head Erase head not seating against tape Dirty erase head Defective oscillator circuit Improper bias frequency
Excessive Hum	Defective head Filter or decoupling capacitor Defective rectifiers Loose or open leads on tape head Poor grounds Bad amplifier stages
Excessive Noise	Noisy transistor Noisy resistors Noisy motor Defective input coupling capacitor Defective integrated circuit Defective cassette Magnetized tape head

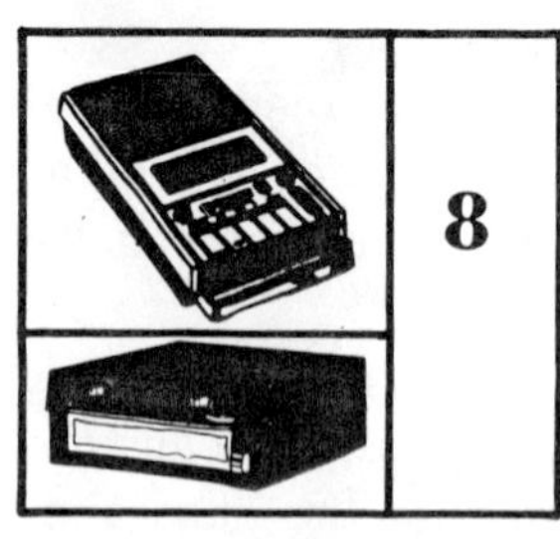

8 Troubleshooting Home & Auto Cassette Players

There are auto cassette players that record *and* play, play only, and others that have integral *AM* and *FM* radios. You will find the stereo circuits in all units similar to the circuits we've been examining up to this point. In general, these units take a bit more time to repair because of their compact construction and more complex mechanical variations.

Direct-coupled transistors are typically found in the preamp stages. The two af or driver stages are either direct- or RC-coupled to push-pull audio output circuits. A transformer is commonly used for coupling between driver and output stages (Fig. 8-1). In most cases the output circuits are transformerless types with electrolytic capacitors used to couple audio to the speakers while preventing dc from leaking through.

The auto cassette player is front slot-loaded to provide safety and easier operation. Generally, these players incorporate cassette ejection. The cassette can be ejected at any time with a manual eject button. When the tape has finished playing, the cassette may be ejected automatically (coupled with automatic shutoff). Some auto cassette players will play both sides of the tape before shutting off the player, while others offer automatic reversing with no shutoff feature.

THE STAAR SYSTEM

Most cassettes are loaded from the top or front of the ordinary cassette player, as shown in Fig. 8-2; you either place the cassette down over the hub spindles or place the cassette in a slotted container and push it into place. With either loading method the hubs and spindles remain stationary. In the auto, the cassette must be loaded from the front. Since the two sprocketed tape reels inside a cassette must slide down over the two spindle hubs another method was used. Mr. Theo Staar, of Belgium, designed a mechanism that permits cassettes to be front slot-loaded. This is known as the Staar system.

When the cassette is inserted in the front slot, the backward motion of the loading lifts the hubs and capstan into position. The mechanism platform with the hubs and capstans are lifted up into the cassette cartridge. At the same time a pressure roller pinches the tape against the capstan drive assembly. From here, you will find the operation is the same as in any ordinary cassette player.

A new Staar MP mechanism has been recently designed to fill a small space in the auto. The new MP slide-in cassette player mechanism is only 1½ in. high, 4 in. deep and 4¼ in. wide. Not only will this small cassette machine fit easily into

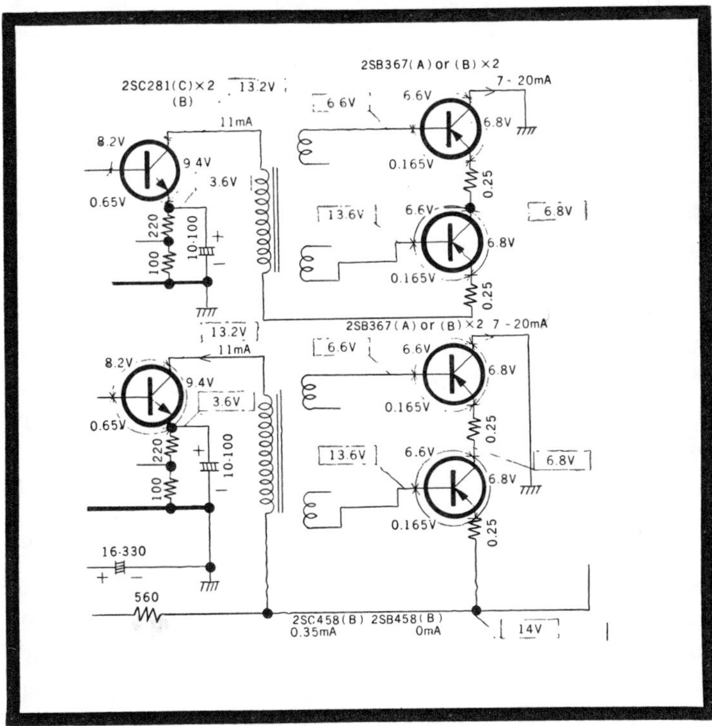

Fig. 8-1. Transformer coupling is found between driver and the output stage. (Courtesy Automatic Radio.)

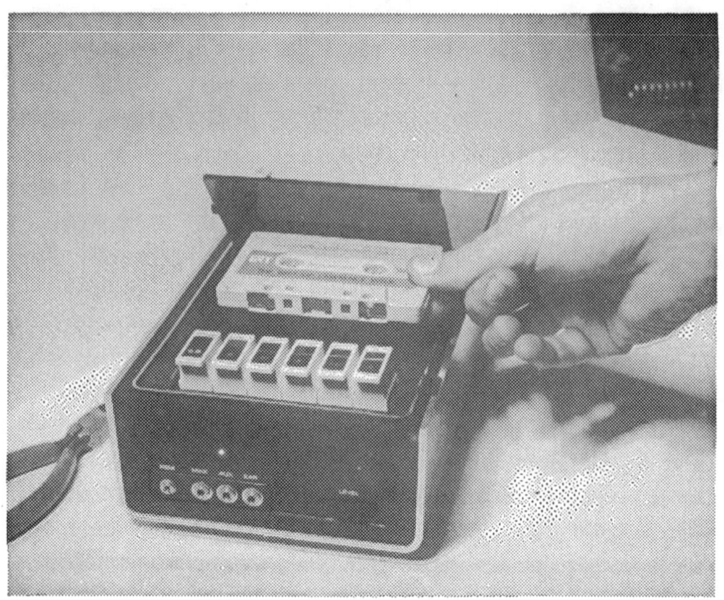

Fig. 8-2. Most cassette players are loaded from the top.

most foreign cars, but the construction is such as to lessen the risk of theft. Besides the United States, sixteen Japanese and foreign countries have been licensed to manufacture the Staar MP system.

The MP mechanism incorporates a simple and effective fast-forward operation. The playback head and pinch rollers are moved away from the tape. A small motor is safely driven at a much higher speed than that of the motor on the conventional player. The fast-forward speed transfers a full C60 cassette reel in approximately 90 seconds.

The MP mechanism is equipped with an automatic stop device which electronically senses the revolution of the takeup reel. The cassette ejects when the end of the tape is reached or the unit is turned off. Any incidental erratic operation that causes a cessation of tape motion will trigger the eject mechanism within one second. This quick cassette ejection prevents damage to the cassette and mechanism.

The automatic reverse play system is accomplished with two flywheels and two capstan drives. Each capstan has its own pressure roller. The capstan drives fit into the two outside holes of the cassette cartridge. One motor drives both capstan and flywheel.

Two latches (instead of one) hold the platform into position. After the first side has played, the sensing switch trips the bottom latch. The reverse pressure roller moves into position and plays the other side of the tape. When both sides have finished playing, the sensing switch activates the solenoid and the top latch trips downward. The cassette ejects and the motor stops rotating.

NO TAPE MOVEMENT

First, check to see if the dial light is on. In case of no dial light or motor noise suspect an open fuse or defective cassette insertion switch. If the player has an AM or FM unit built in, see if the switch is on *tape* or *radio* position. If the radio is working, you know the cassette switch may be open. Check to see if the eject solenoid makes a noise (see schematic, Fig. 8-3). This will prove that power is getting to the tape player. Don't overlook an open fuse in the power lead.

If you hear the motor running and still no tape movement, suspect a broken or loose drive belt. Check for a dry or binding spindle assembly, The idler wheel may be deformed, worn, or frozen. A worn or binding idler wheel will cause erratic takeup operation. Suspect a defective idler

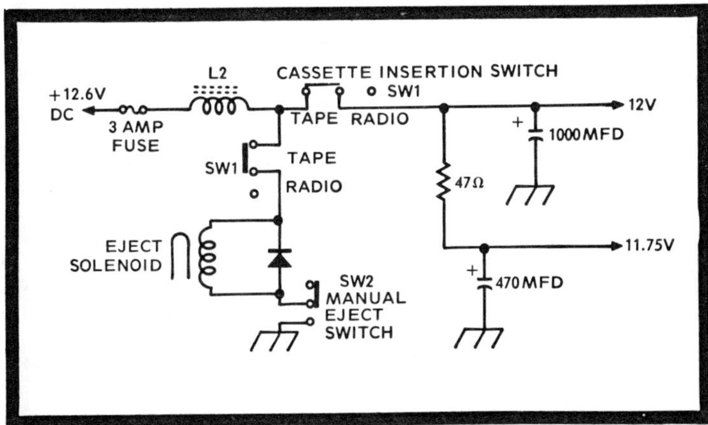

Fig. 8-3. You can easily check for power operation by pushing manual eject SW2 and listen for eject solenoid noise.

spring and bent arm. Also check for a frozen capstan—flywheel assembly.

In case the motor does not rotate when power is applied suspect an open or dry motor. Remove the drive belt and trip the cassette switch. See if the motor takes off and appears noisy. Measure the dc voltage at the motor terminals and check the continuity of the motor windings. Either the motor or a possible electronic motor speed regulator circuit is defective.

NO FAST FORWARD

Fast forward is accomplished in the auto cassette player with idler wheels or a change of motor speed. If the cassette operates normally in *play* but appears erratic at times or has no fast-forward operation, suspect a defective idler. The idler may be dirty and sticky with grease, or it may be bent out of alignment. Check to see if the idler is deformed or binding. Also check the fast-forward idler arm (it too may be bent out of position).

In auto cassette players with fast-forward motor speed, check the fast-forward switch. If the player will rewind but there is no *fast-forward* or *play* operation, the switch may be

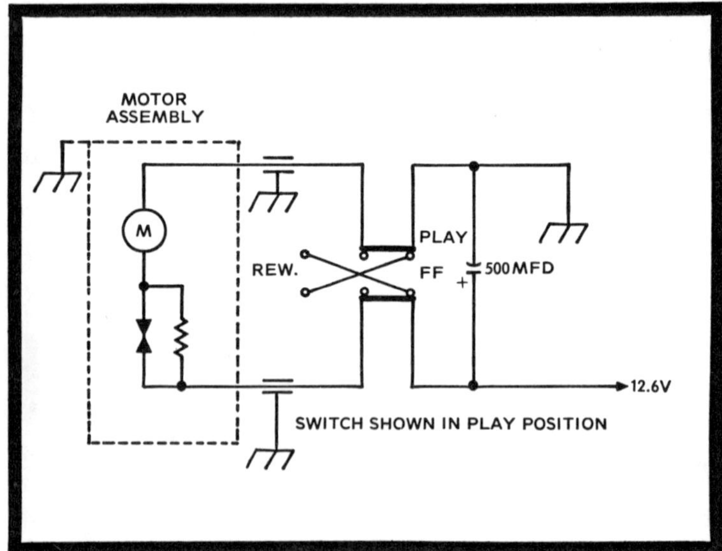

Fig. 8-4. Check for open or dirty switch.

dirty or open (Fig. 8-4). In this type of circuit the *fast-forward* and *play* speeds share a common reel-drive source. Of course, in fast-forward position another idler wheel drive train increases the fast-forward speed. Don't overlook a defective high-speed winding within the motor.

CASSETTES KEEP EJECTING

The auto cassette manual ejection circuit (Fig. 8-5) is similar to that of the auto 8-track cartridge player. In automatic eject operation a sensor switch is connected to the supply spindle and when rotating the cassette will continue to play. If the supply spindle or sensor stops, the cassette is ejected. The cassette is also ejected at the end of the tape reel.

Most problems found with automatic cassette ejection mechamisms are the cassette ejects too soon and plays for only a few seconds. Quick ejection can be caused by a sticky cassette. If the cassette will not unwind properly or become dry in reel mount, the cassette will be ejected. Try a new one to see if the cassette is defective or if the problem lies in the player.

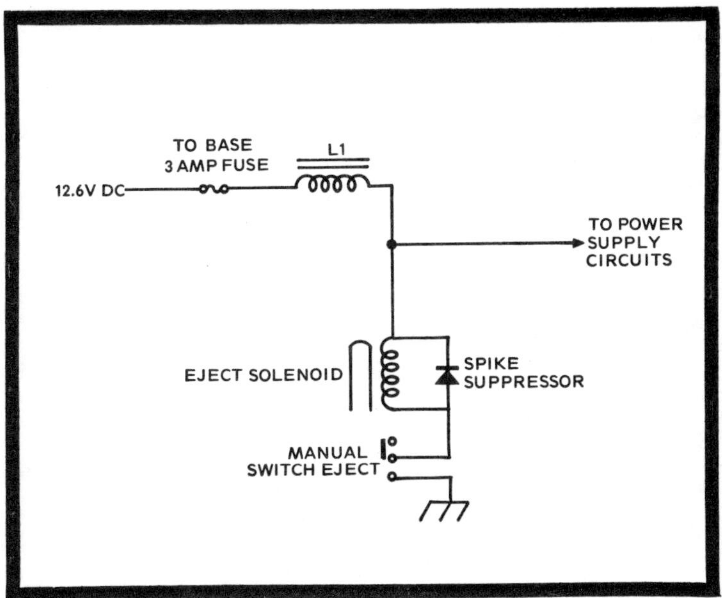

Fig. 8-5. The auto cassette manual eject circuitry.

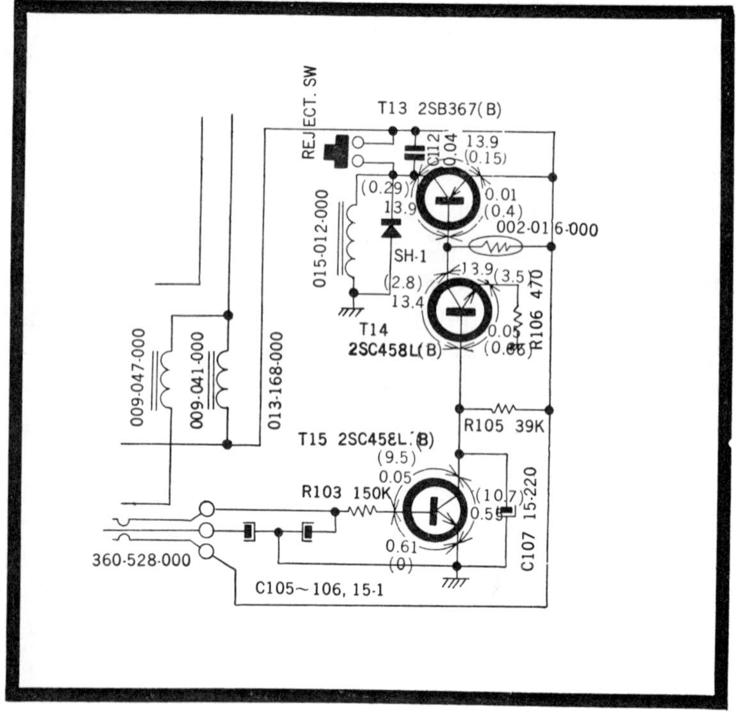

Fig. 8-6. A typical pulse sensor cassette ejection circuit. (Courtesy Automatic Radio.)

Interrupted power will also eject the cassette. A dirty or defective sensor leaf switch and commutator will produce erratic cassette ejection. Most early cassette ejection is caused by dirty leaf switches. Simply clean up with alcohol.

If the cassette fails to latch in the player, suspect a worn or binding latch assembly. A worn or deformed back roller may not let the subchassis latch hold in place. Check for a weak solenoid spring. The solenoid core or lever may be binding and will not let the cassette seat properly. Don't overlook a weak ejector spring. In many cases you can cure the problem by adjusting the tension of the ejector spring.

A typical pulse sensor cassette ejection circuit is shown in an *Automatic Radio* ACS-6000 (Fig. 8-6). A unique SCR type detection circuit of a cassette ejection system is found in the *Realistic* cassette player (Fig. 8-7). You will find no transistors in this circuit; only resistors, capacitors, diodes, and SCRs control the ejection solenoid. You may eject the cassette manually with switch SW6.

WEAK LEFT CHANNEL

You can signal-trace the weak or defective stage and compare it with the good channel. When using an audio signal generator always keep the signal just low enough to be workable. Besides the audio signal generator you may use the speaker and scope as indicators.

Since parts are jammed together in the auto tape player, it is difficult to locate the various stages. How are you going to locate the weak stage without circuit diagram and service information? You know that the left channel is weak, but where is it located? Actually, start at the speaker output

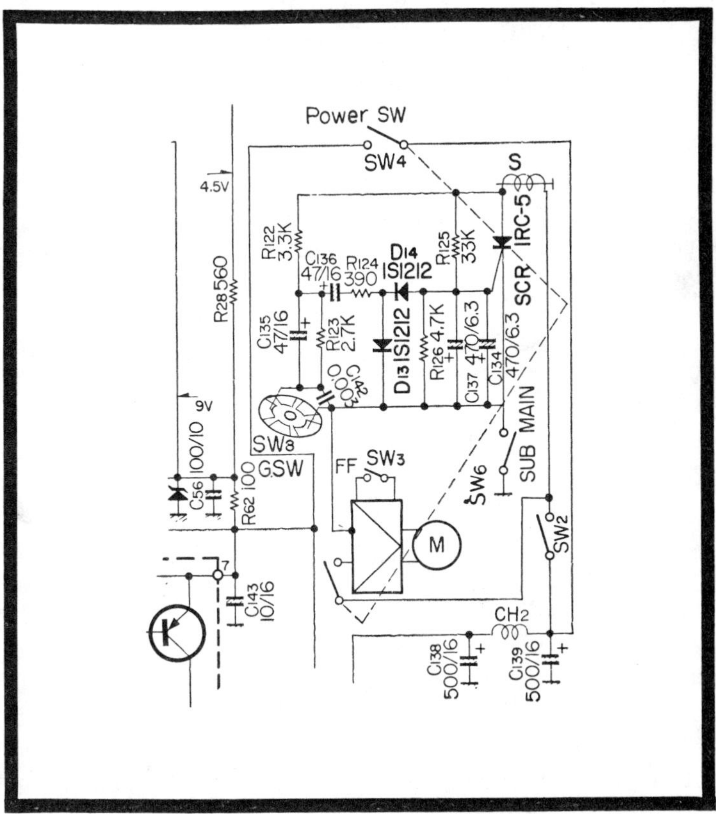

Fig. 8-7. SCR ejection circuit found in a Realistic Model 12-1825. (Courtesy Allied/Radio Shack.)

terminals with a scope. Insert a test or music cassette and turn up the good channel.

Now, begin at the output transistor. Touch the scope lead to metal collector terminal of the output transistors. Generally, one power transistor of each channel is isolated above chassis ground. Stay with the output transistor that indicates a signal. Rotate the volume control up and down to be sure. You will find the corresponding output transistor is right alongside. If in doubt, check both identifying numbers on the transistors. You can locate all working transistors with this method. Now do the same with the weak channel and compare the signals. Where the signal becomes weak you have located the defective stage.

ADJUSTMENT OF PLAYBACK AMPLIFIER GAIN

In some auto cassette players, small semifixed variable resistors are used to preset amplifier gain. With this method each channel is correctly adjusted for the same amount of gain. You will find, when the audio circuits have added a little age, one side may be a little lower in gain than the other. You can adjust these preset gain controls to make up for good overall balance.

For this adjustment, use a test tape in the vicinity of 300 to 500 Hz. Use the correct load resistance to match the unit's requirements (4–8 ohms). Set the balance control on dead center. Adjust both output terminals for exact voltage readings across the load. Depending upon the power output of the player, the voltage will vary from 0.5 to 2.5 volts. Some cassette players have a side hole in the case for amplifier gain adjustment. In others you should make this adjustment before replacing the metal covers.

ALC PROBLEMS

Low recording level problems can occur in the bias oscillator. ALC (automatic level control), or amplifier circuit.

When the microphone signal is too great, the ALC circuit reduces the audio as not to overload or distort the recording. In the auto, ALC is a vital circuit since there are disturbing outside noisy conditions—the motor and wind conditions.

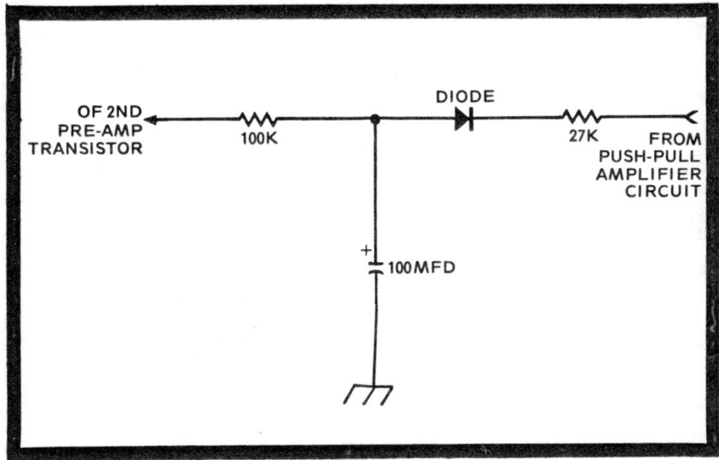

Fig. 8-8. A simple diode detection type ALC circuit of a small cassette player.

In most cassette ALC circuits the output is tapped at the push-pull output stages and fed to the second preamp circuit. You may find only a diode, corresponding resistor, and electrolytic capacitor in the ALC circuit of a small cassette player (Fig. 8-8). In auto and stereo cassette tape decks you will locate a diode doubler with one or two transistors in the ALC circuit (Fig. 8-9).

When ALC recording is not obtained, suspect a defective ALC transistor. Check the single or double diodes for leaky or

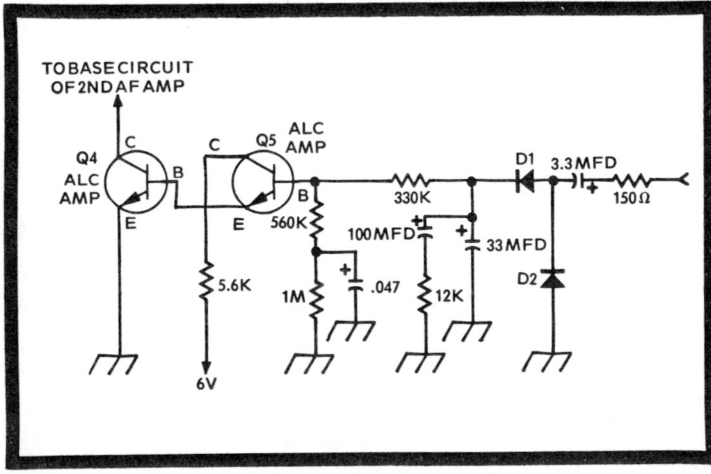

Fig. 8-9. A diode-doubler transistor ALC circuit.

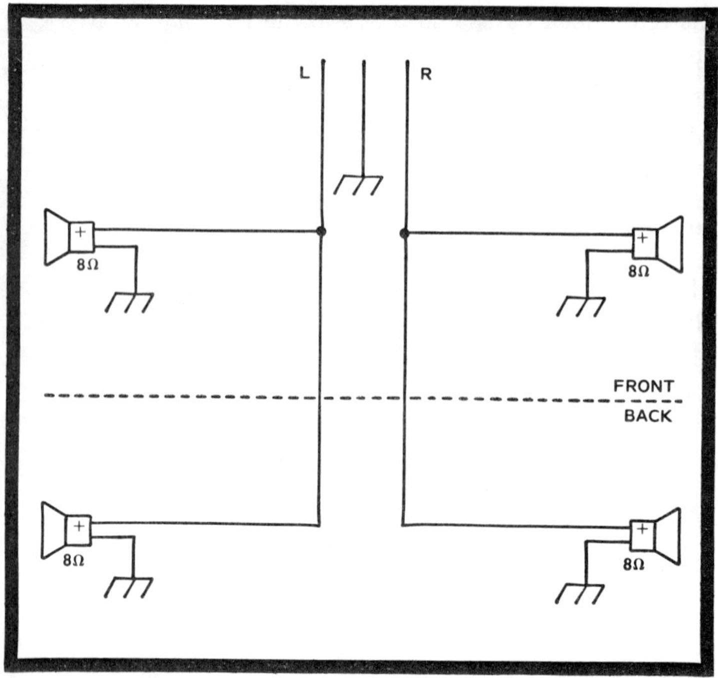

Fig. 8-10. A pair of 8-ohm speakers may be paralleled to create a 4-ohm load. With one such pair on the left and another on the right, the in-car stereo effect may be enhanced considerably.

open conditions. Remove one end of the diode from the PC board for correct resistance tests. With one end of the diodes out of the board, check corresponding resistors for correct value. Most problems found in the ALC circuits are traceable to transistors, diodes, and electrolytic capacitors.

AUTO STEREO SPEAKER HOOKUP

Most auto cassette and 8-track players are designed to power 2 or 4 speakers; the best performance is achieved with the use of matched speakers. Mixing speakers with different impedance values can cause a serious mismatch in the amplifier. The results range from poor volume and tone quality to damaged transistors.

You should always install a minimum of two speakers, one on the right side of the car and one on the left to get the proper stereo effect. For the ultimate in stereo performance, install a set of 4- or 8-ohm speakers, two in front and two in the rear. Hook up two speakers to the left channel and two speakers to the right channel, as shown in Fig. 8-10.

An interesting variation of the 4-speaker hookup is that of Fig. 8-11. This arrangement essentially converts the player to a 4-channel "synthesis" hookup that can effectively decode Sansui, Dynaco, and Electro-Voice 4-channel matrixed stereo records.

The two front speakers can be mounted in the bottom side of the two front doors, as far forward as possible. Sometimes the speakers can be mounted in the front kick panels. You should double-check either location for proper speaker clearance. Use narrow flush mounted speakers in the doors;

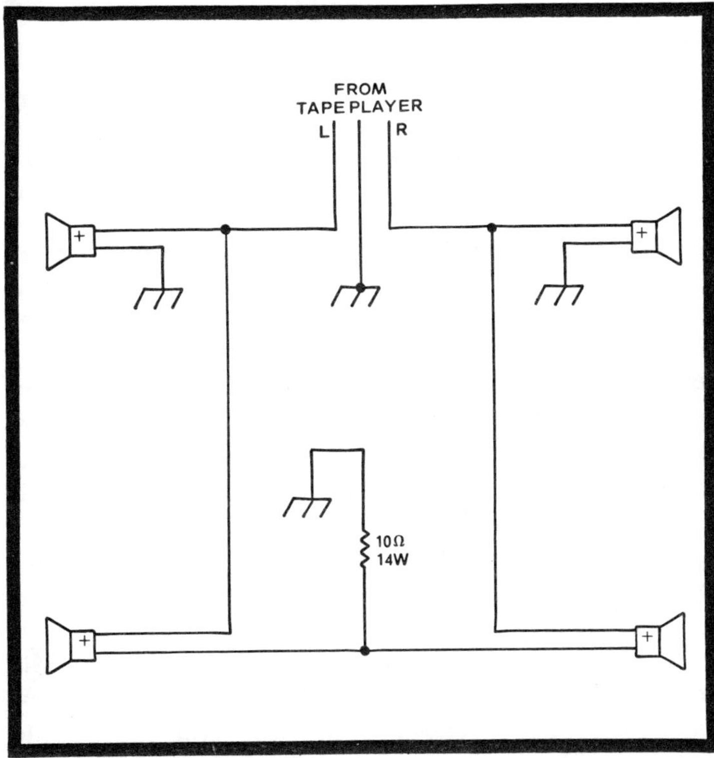

Fig. 8-11. This simple speaker arrangement allows a conventional stereo player to be used as a 4-channel setup. The 10-ohm resistor matrixes the rear-channel signals to create a concert-hall effect. Again, only speakers of 8 ohms or more can be used with this circuit; paralleling low-impedance (4-ohm) speakers results in heavy loading, mismatch, and probable damage to some amplifiers at high volume levels.

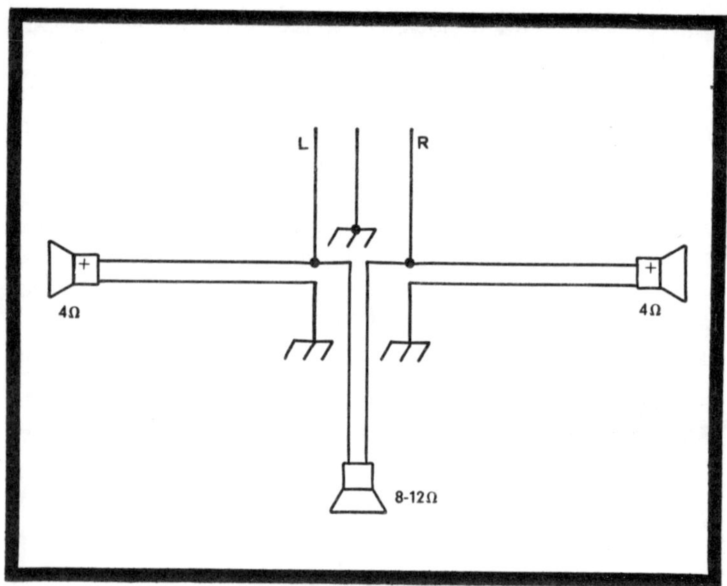

Fig. 8-12. This circuit will closely approximate the sound obtainable with the 4-channel arrangement. Some left—right separation is sacrificed in favor of an exceptional amount of separation between front and back.

otherwise, the window-mechanism will be thwarted. Sometimes the small surface mounted speakers can be hung at the top of the side panels. Make sure the emergency brake pedal and auto lid release mechanism will clear the speaker mountings.

At the rear, mount the speakers in the rear deck. Most new cars have speaker cutouts on both sides of the rear deck. You can just cut through the cardboard of rear deck for easy mounting. Use an ice pick or a sharp tool and punch out the speaker mounting holes.

If there is only one central speaker cutout on the rear decks you can use it effectively in a 3-channel arrangement that is almost as impressive as the 4-channel version already described. The arrangement (Fig. 8-12), requires two matched speakers plus another of up to twice their impedance. The only drawback to this arrangement is an acoustic cancellation of low-frequency information with some stereo tapes. A polarity-reversing switch for the rear speaker will negate this problem, however. The operator simply switches the speaker for best bass reproduction. (See Fig. 8-13.)

In most new autos, you'll run into uneven surfaces for mounting speakers. If there is a gap at any given area around the speaker fill in this area with General Electric silicone rubber cement. Simply spread it around the speaker enclosure until all air leaks are completely sealed, and you have a professional mounting installation.

When grounding speakers to the car metal (for common ground), use a star washer to bite into the painted metal. Sometimes it is best to run two wires to each speaker to be sure of a good ground return. Run the speaker wires behind the front kick panel at the bottom of the front doors. Remove the light metal kick plates from the door openings and place the wire in the available groove under these panels. The back

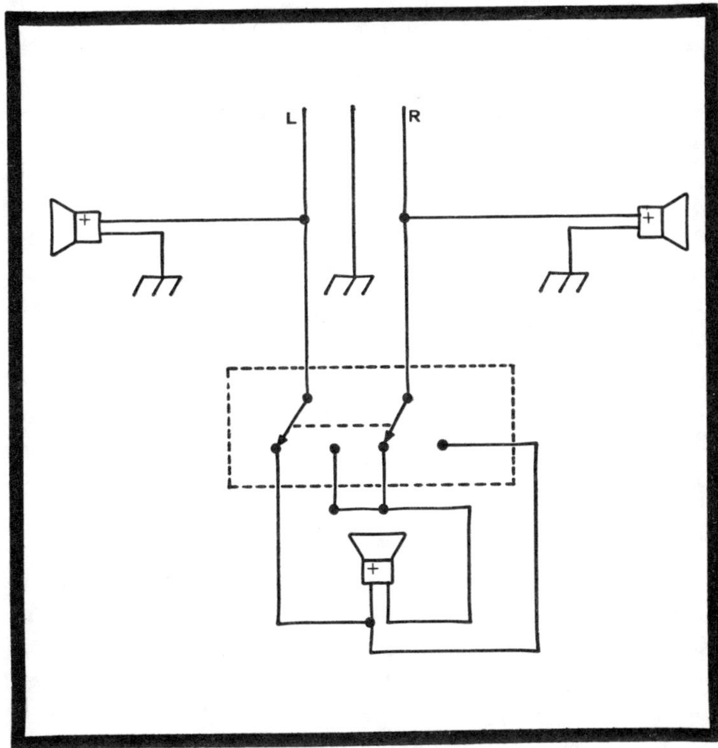

Fig. 8-13. Polarity-reversing switch for rear speaker allows enhancement of bass notes without regard to program material or encoding system used.

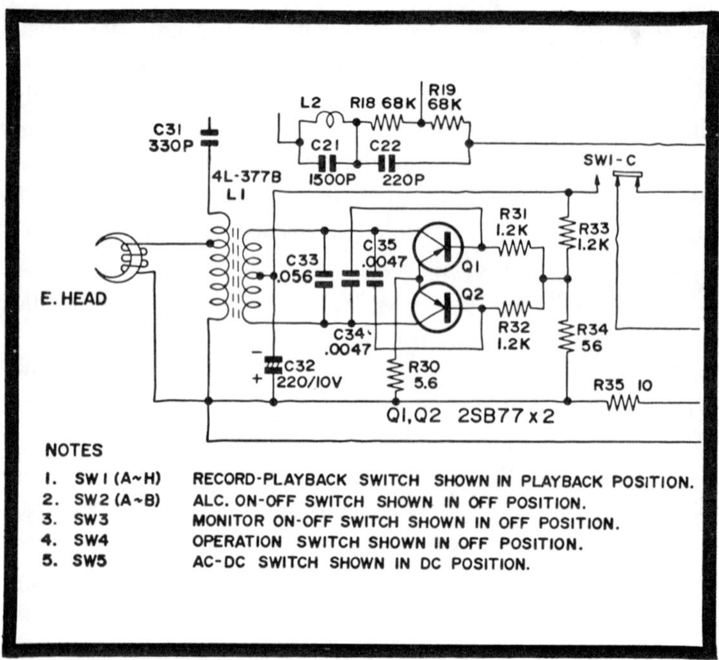

Fig. 8-14. A push-pull bias oscillator circuit found in a stereo cassette player. (Courtesy Sharp Electronics.)

seat may have to be removed to extend speaker wires into the trunk area. In most cars a metal or cardboard back should be drilled to extend speaker wires into the trunk area. Always solder or use wire connectors in attaching wires to the speaker terminals.

A very important rule: Never use a wire gage less than 18 AWG for rear-deck installation. Thinner wires are more apt to trade their signals for noises and other radiation pickup.

SERVICING CASSETTE STEREO TAPE DECKS

Professional players normally include two separate VU level meters. (Advent's Model 201 is an exception, with its single VU meter.) In some stereo players FM multiplex (stereo) radio circuitry is built in; in others you may also find a stereo turntable.

Most cassette stereo player—recorders have a separate erase and stereo tape head. A push-pull bias oscillator circuit (Fig. 8-14) is found in most models. The cassette tape deck may have an electronic cassette ejector system. A ripple filter transistor circuit is found in the power supply of some players (Fig. 8-15), while others incorporate conventional bridge rectification for the power supply voltage source.

The cassette stereo tape deck may consist of only a preamp audio circuit and a nonamplifying emitter-follower stage (for impedance matching). Figure 8-16 is a 3-panel schematic that shows a complete tape deck (*Realistic* 8CT-5). The preamp stages (first panel) are direct-coupled *npn* transistors.

The audio amplifiers (second panel) are fed through series coupling capacitors C12 and C112. The third panel shows power supply, audio output, and level control circuits. While not all decks use photocells like the Realistic, this schematic is typical of modern solid-state design, and should help you troubleshoot a similar unit.

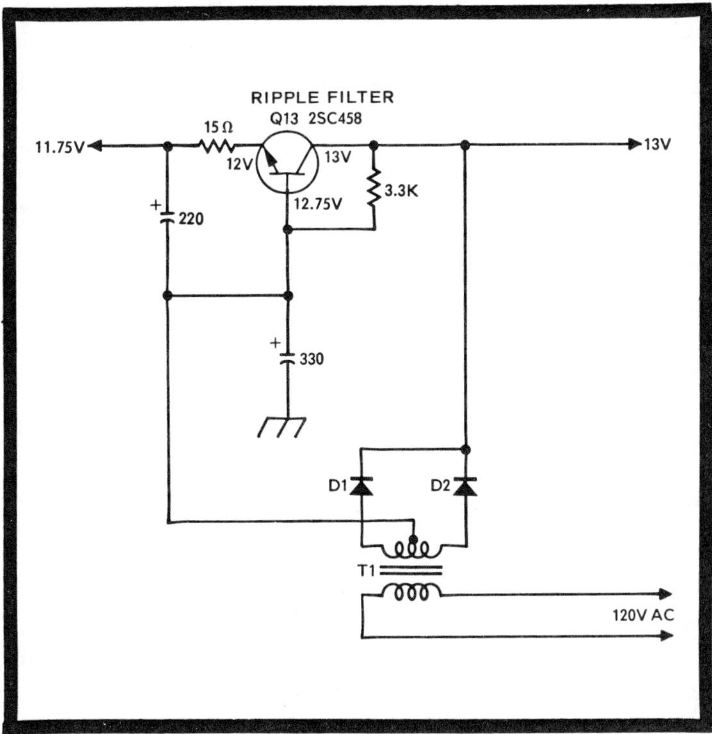

Fig. 8-15. A ripple filter transistor circuit found in some power supplies of cassette tape decks.

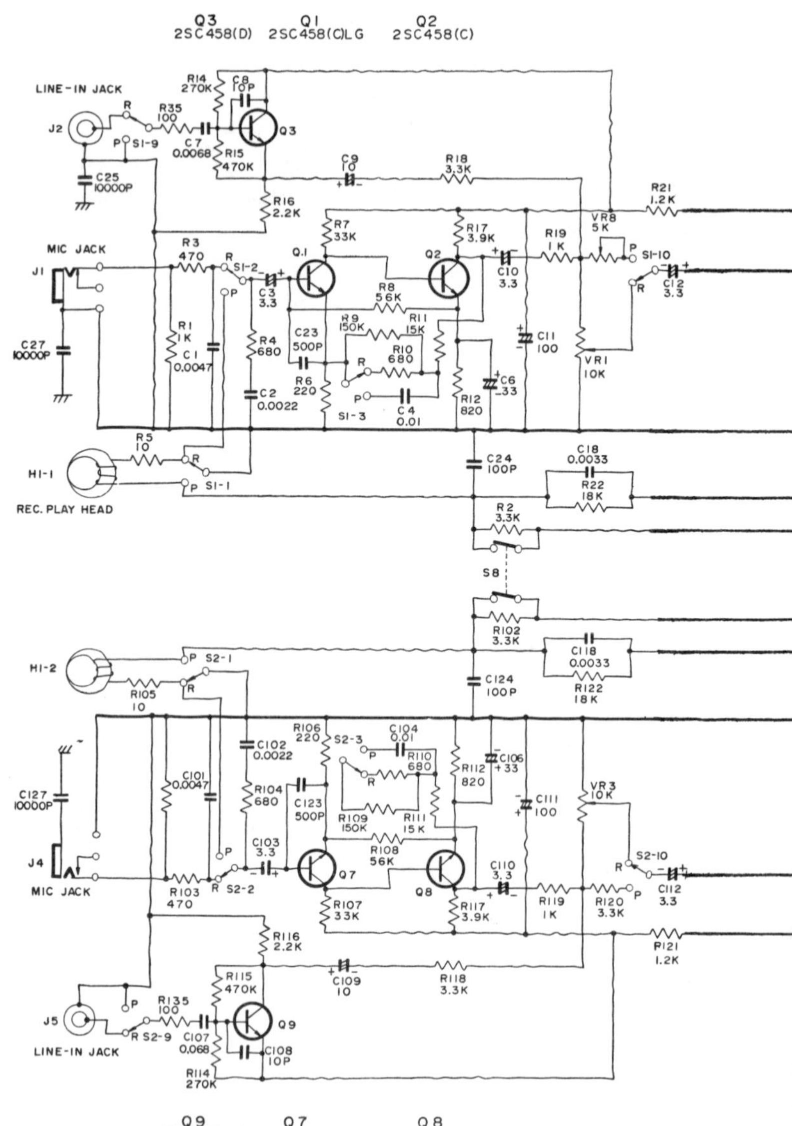

Fig. 8-16. A typical preamp circuit of a stereo cassette tape player.

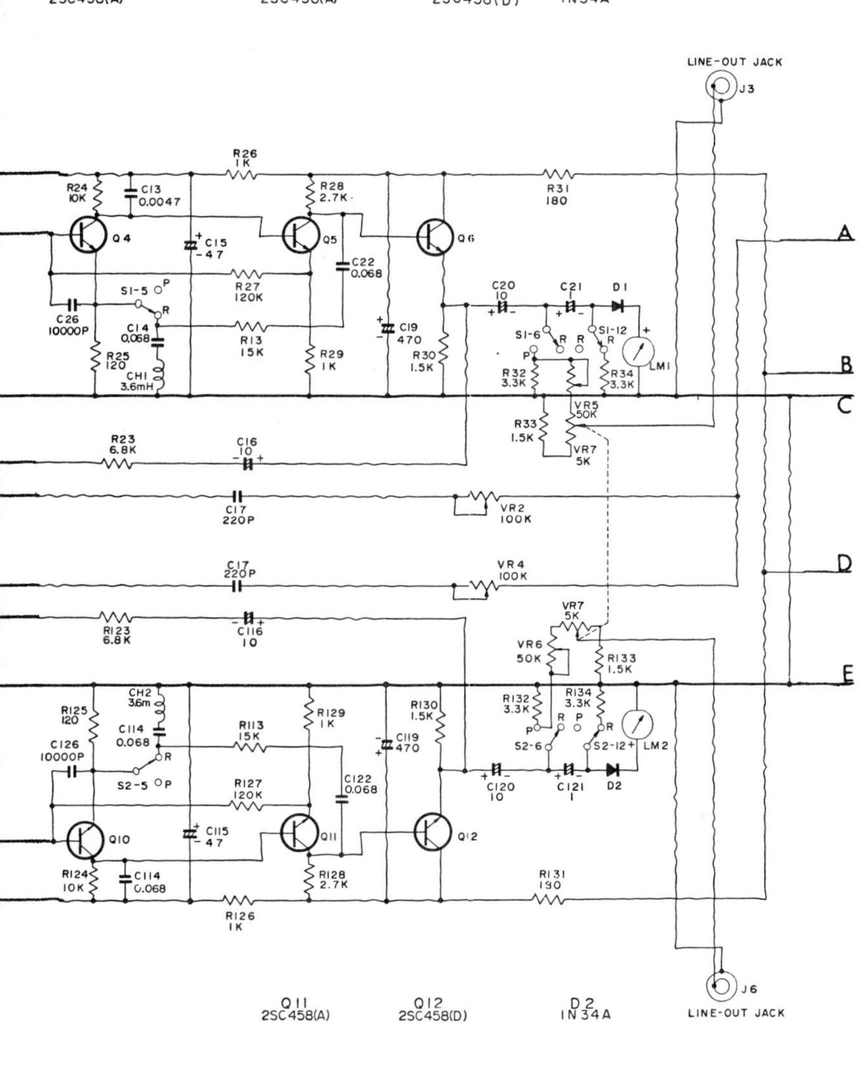

Fig. 8-16. Con't. on 180.

Q13, Q14
2SB77(CP)X2

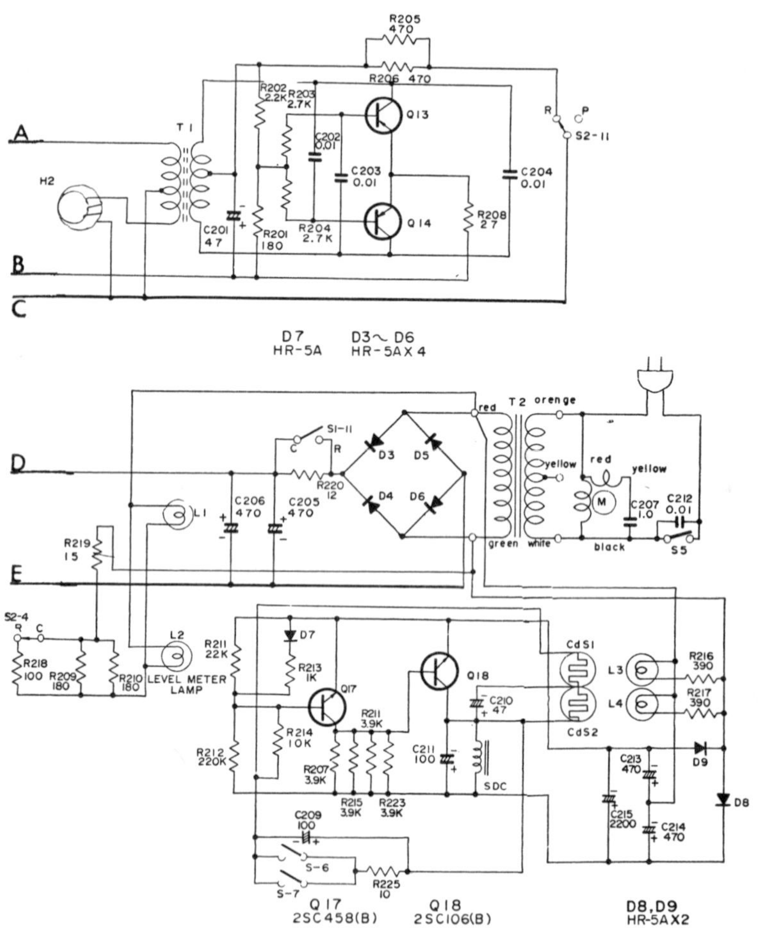

Fig. 8-16. Con't.

SERVICE BY MEASUREMENTS

You can check the operation of the bias oscillator stage by setting up the test instruments as shown in Fig. 8-17. With this hookup, you can set the amount of oscillator bias, check the frequency, and test for proper operation. Generally, the oscillator voltage and frequency are given in the manufacturer's literature.

1. Switch the cassette player to *record*.
2. Connect VTVM to test point.
3. Adjust variable oscillator bias resistors (Fig. 8-18).
4. Adjust oscillator coil at 53.5 kHz.
5. After setting the bias frequency of the oscillator, repeat step 3.
6. Make oscillator bias adjustments of both stereo tape heads.

You can check the operation of the erase head with a test instrument setup. Connect the VTVM and scope as shown in Fig. 8-19. The manufacturer may give a different method or value of R1, but the erase voltage check is the same. If you suspect distortion and improper tape erasing, try this method of detection.

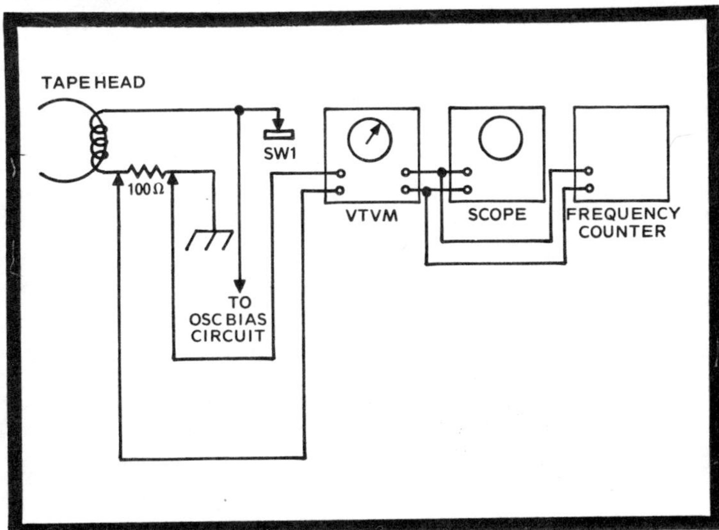

Fig. 8-17. A simple bias oscillator frequency adjustment with hookup. (Check each stereo head section in this manner.)

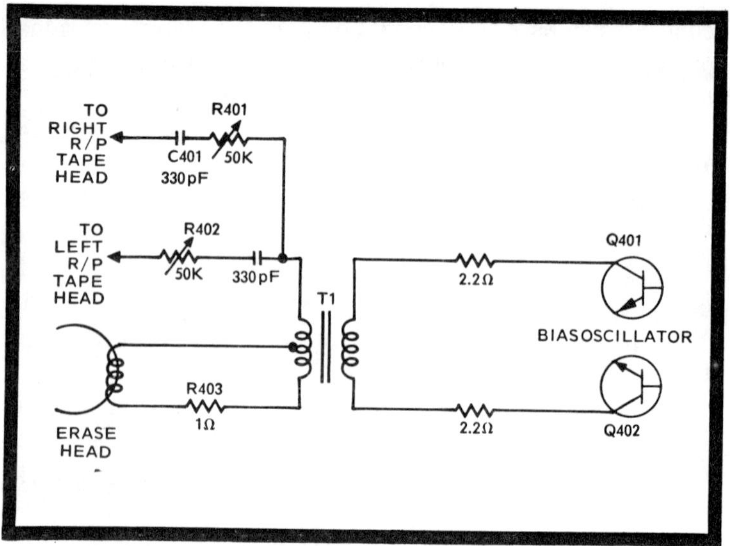

Fig. 8-18. Adjust R401 and R402 for correct bias to right and left gaps of tape head.

1. Switch the cassette player to *record*.
2. Connect VTVM across R1.
3. Erase voltage is normal if VTVM reading is over 0.05V (50 mV).

QUICK METHOD TO ISOLATE DEFECTIVE STEREO STAGE

You can quickly isolate a weak or dead audio stage in a stereo amplifier by injecting signal from a pencil noise generator. This method applies only to high-gain audio stages, and especially those with metal-case transistors. Turn both volume controls wide open. With Dolby off, you will hear a transistor rushing noise in the good stage.

Now start at the af or driver transistors and work towards the front of both stereo amplifiers. Touch the metal probe of the signal generator to the metal top of the af or driver transistors. If the signal level is about the same move on to the next transistor in each stereo channel. You can inject signal at the metal end of the electrolytic coupling capacitors. As you move towards the preamp stages the volume will get louder. You may have to turn the volume control down as you proceed. Compare the volume or signal of the defective channel with the good one. When the signal level stops or lowers you have located the defective stage.

WHAT TO DO WHEN TRANSISTORS "POP ON"

While working on an audio stage, the defective transistor "pops" on. What should you do in this case? Generally, the transistor that was touched with the voltmeter probe or while checking with an in-circuit transistor tester is the culprit. You will find in many cases a transistor will "pop" on when touched. But before removing it, it is best to be sure.

First spray on several coats of *cold spray* (available at local electronic parts distributor) on the suspect transistor. Cool the transistor at least three times. If this doesn't work, spray the other transistors in the defective audio section. Turn the player off and let set for a few hours and try it again. The last resort is to replace the transistor that "popped on."

ADDING OTHER PROBLEMS

When servicing radios and amplifiers sometimes the dial cord is right in the way of the suspect transistor, so when you try to remove the transistor leads from the PC board, you unwind or break the dial cord. It is quite discouraging when

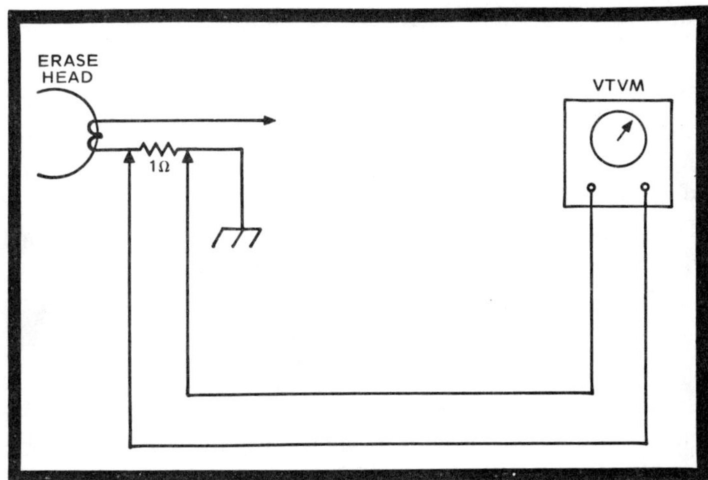

Fig. 8-19. How to hook up and check out the operation of the erase head.

the soldering iron accidentally touches the dial cord and it pops off of every pulley. Besides replacing the defective transistor, you have to install a new dial cord. And dial cord stringing can be very complicated if you don't know which pulley it's routed to first.

Protect that dial cord initially by applying masking tape or a cardboard tube carton over the dial cord area. Slit the tube or transistor carton, place it over the dial string, and fasten the ends together with masking tape. Slide the carton near the soldering area and go to work. It's also wise to examine the cord routing and draw it on paper in advance if there's a chance of breakage. After transistor replacement you can easily remove the carton and prevent adding another problem to the original one.

PREVENT MARRING CASSETTE FRONT PIECE

To prevent scratching or marring the front plastic or polished metal pieces of a cassette player, apply masking tape. Place a layer of masking tape over small polished metal areas such as flip-up cassette cartridge holders. Cover the larger areas with rubber or fiber packing material. In many instances, you have to turn over or handle the player when servicing it, and before you know it you have added a few more scratches. Of course, the cassette player should always be placed upon a mat or service cloth before attempting to repair it.

STEREO CASSETTE DECK TROUBLE CHART

TROUBLE	CAUSE
Automatic ejection circuit does not function	Check driver and solenoid transistor for open condition
	Shorted driver transistor
	Broken PC board
	Open emitter resistor
	Shorted thermistor
	Defective plunger
Cassette ejects when inserted	Open or shorted transistor
	Open or broken emitter and bias resistor
	Broken PC board

Ejects too quickly	Bad contacts on commutator switch Bad lead on commutator switch Supply or takeup reel binding Tape snags or pulls
Cassette does not eject	Open or defective solenoid coil Shorted suppressor diode across solenoid coil Poor contact of switching mechanism Bent or broken plunger lever
Recording system disabled	Slide switch defective Poor power supply voltage Defective recording amplifier Bad contact switch
Distorted recording	Defective oscillator transistor Improper bias (too much, not enough) Open oscillator coil Open input capacitor Defective erase head Defective ALC circuit
Low or weak recording	Poor bias current Defective tape head Shorted ALC transistor Improper adjustment of internal gain control
No bias oscillation	Defective oscillator transistor Broken or open oscillator coil Improper oscillator supply voltage Open or leaky filter capacitor (oscillator circuit)
No erase	Defective erase head Poor or open leads on erase head Improper alignment of erase head Erase magnet broken or missing

Symptom	Possible cause
Tape speed increases	Insufficient pinch roller pressure Excessive takeup torque Capstan and pinch roller out of line
Slow fast-forward	Insufficient roller pressure Insufficient or excessive intermediate idler pressure
Tape is not taken up (bunches)	Broken idler belt Defective drive pin assembly Warped takeup hub Defective cassette
Slow tape speed	Loose motor drive belt Warped supply hub Defective motor Defective cassette (wound too tight?) Oil on motor belt and drive mechanism
Excessive wow conditions	Loose motor belt Bent or worn capstan Oil on motor belt Foreign matter on flywheel Twisted motor belt Frayed drive belt Dry capstan bearing
Poor or slow rewind	Brake will not release Improper pressure of intermediate idler Improper pressure of right idler roller assembly Excessive pressure of intermediate idler Excessive pressure of right idler roller assembly
Noisy operation	Warped and noisy cassette Noisy motor Defective motor bearings (Inferior cassettes will squeak when humidity increases)

Servicing the Auto Cartridge Player

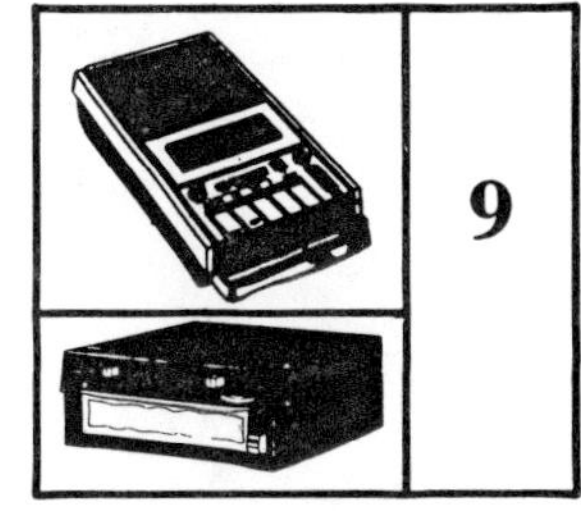
9

Since most "wheel-equipped" teenagers have stereo players in their cars, you will naturally receive more of these units to repair. Most problems associated with other tape players can be found in these as well. Let's take a look at some common cartridge-player problems.

NO AMPLIFIER HISS—NO TAPE MOTION

No tape motion and amplifier hiss indicates that no power is getting to the tape player. Trigger the cartridge insertion switch with a pencil through the tape cartridge opening. See if the capstan–flywheel begins to revolve. If not, notice if belt is off and the motor pulley is turning (Fig. 9-1). In some units you can see the motor pulley rotate through the cartridge opening. Notice if any program indicator lights are on.

If there is no tape motion, check first for an open fuse in the power lead. And check the wiring socket connections. If these preliminary measures lead nowhere, remove the covers and check for a defective cartridge switch. Short across the cartridge switch with a pair of alligator clips and lead. In case the tape player starts up, replace the defective cartridge switch (or repair it by carefully rebending the leads). Check the wiring from the external socket to the switch. Look for burned or open PC board wiring. In some units, power is furnished through the PC wiring to the motor and program lights.

Don't forget to check the choke coil in the power circuit for open windings. Some of these coils are air-core types and others have iron-core centers. In some models, a bypass capacitor is found from the hot side to chassis ground. You will find a blown fuse if the bypass capacitor shorts or if the

Fig. 9-1. See if the belt is off and motor pulley is rotating.

suppressor diode across the program solenoid shorts internally.

With no amplifier hiss or capstan action, suspect no power voltage applied to the amplifier section. Check the power supply voltage to the amplifier circuits. Generally, you will find one channel dead—not both. The dead or weak amplifier section should be checked with an audio signal generator. You can inject a signal from the audio generator at each stage to uncover the defective stage.

SOMETIMES SLOWS DOWN

Several problems can exist when the tape player slows down in operation. This can be improper setting or alignment of the cartridge in the tape holder. Try to pull the cartridge back and see if the tape returns to normal speed. Move the cartridge from side to side and notice a change of speed. The cartridge tension against the capstan may be too great.

A dry or gummed up capstan bearing will produce abnormal speed (Fig. 9-2). The cartridge pressure against the capstan drive may slow the tape speed. First, clean up capstan–flywheel drive and bearings, then lubricate them. Install a new motor drive belt. These drive belts may often *appear* to have good tension, although they still slip and cause slow speeds.

One of the most common causes of sluggish operation is the cartridge itself—and this is usually traceable to an improper tape-repair job. Some customers may try to wind their own cartridges, which work fine for a play or two but no longer. If this is the case, inform your customer that cartridge tapes are specially lubricated to allow one-reel operation, and the use of regular reel-to-reel tapes just can't be effected without problems.

Let the tape player operate for several hours and periodically check the tape speed. After the tape player has been operating for several hours, apply pressure against the tape cartridge. Notice if tape slows down. Of course, excessive pressure will slow down any tape player, but always check out the stereo 8 player in this manner. This extra outside pressure will show up a slipping drive belt or sluggish capstan drive assembly.

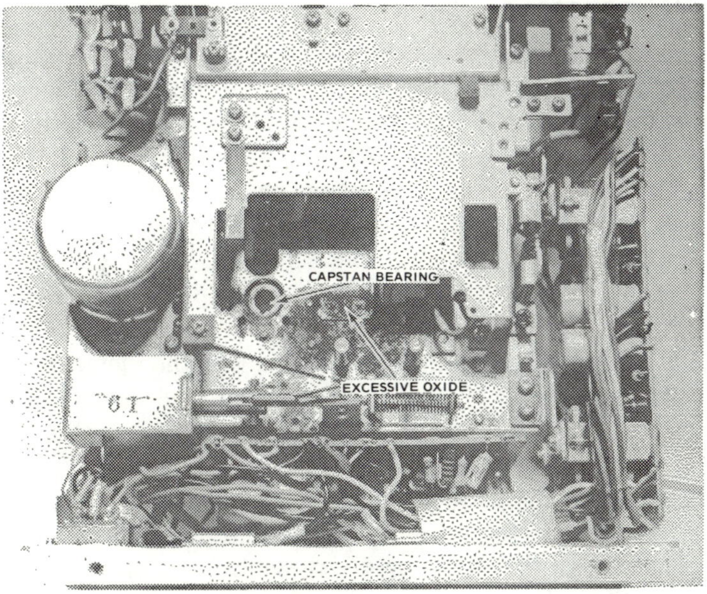

Fig. 9-2. A dry or gummed up capstan bearing will produce abnormal speeds.

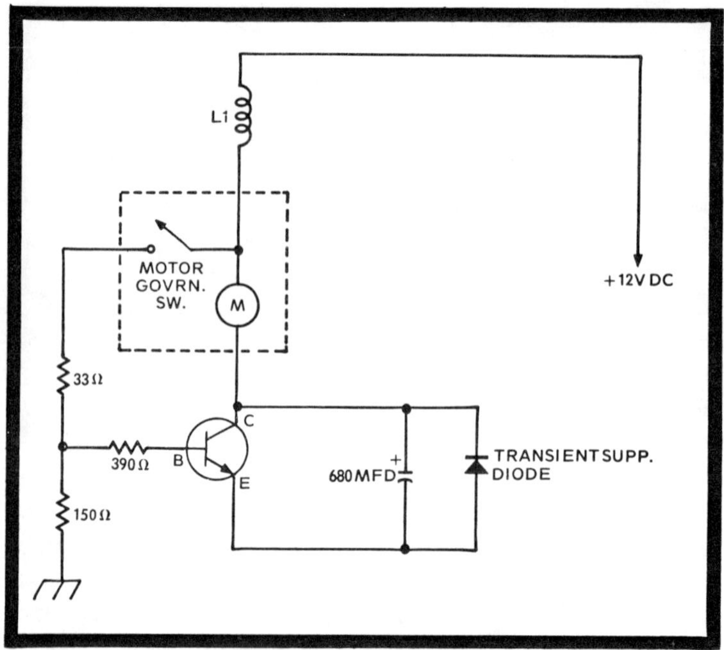

Fig. 9-3. Typical fixed motor regulator control circuit.

STEREO 8 MOTOR CIRCUITS

A typical electronic controlled motor circuit is shown in Fig. 9-3. The motor has an internal governor switch for controlling speed. A power transistor is located in the ground leg of the motor winding. When the motor speeds up the centrifugal switch will close and change the bias of the transistor, lowering or stabilizing the motor speed.

The two-transistor electronic control circuit may be fixed (Fig. 9-4) or variable (9-5). The fixed electronic control circuit is very simple and employs only a few components. The two transistors are direct-coupled with the motor in the collector circuit of the speed control transistor. Most transistors found in the output stage are "power" transistors, larger than other circuit transistors and mounted on husky heat-dissipating metal surfaces.

In the variable electronic speed circuit, the motor is connected to the +12V source through L2 (Fig. 9-5). The ground leg of the motor is connected through L1, a 10-ohm resistor, and speed control transistor Q2. By rotating speed

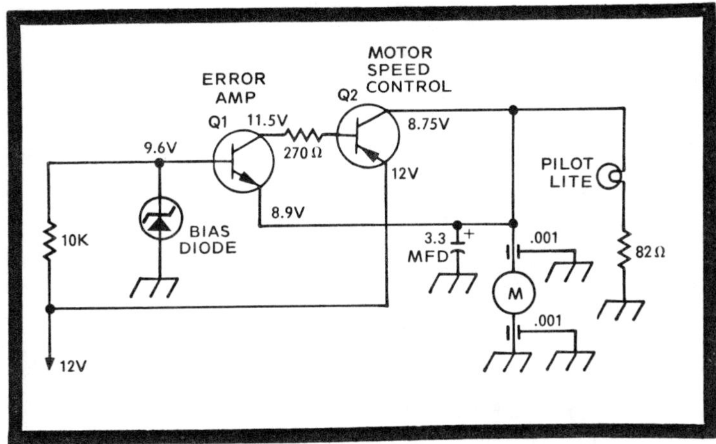

Fig. 9-4. Fixed two-transistor speed control circuit.

control R1, base bias is changed on the base circuit of Q1. The variable voltage conduction of Q1 is tied directly to the base of the speed control transistor (Q2). A small change of voltage on the base terminal of Q2 will vary the resistance path through the transistor, thus changing the motor speed.

A typical three-transistor variable-speed electronic motor circuit is in Fig. 9-6. In this speed circuit, all three transistors are direct-coupled. When rotating the speed control, the base voltage is changed on the dc motor amplifiers. The voltage change on this base terminal of Q3 determines the motor speed. Transistor Q3 is a conventional power amplifier.

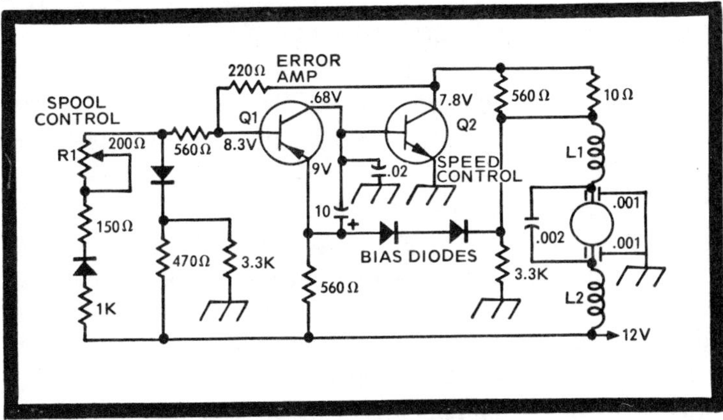

Fig. 9-5. Two-transistor variable motor speed control.

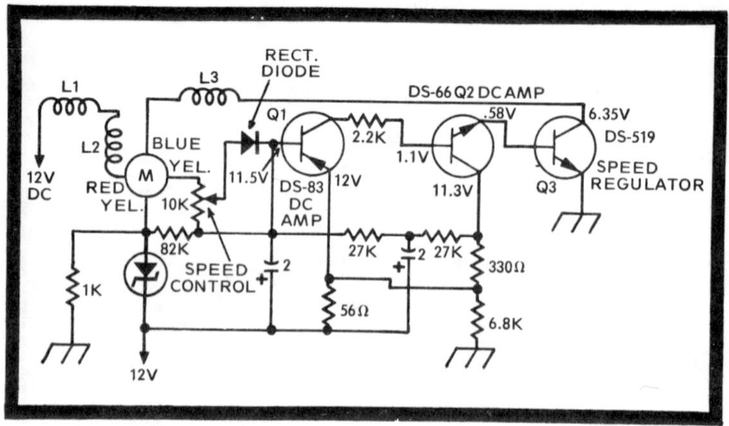

Fig. 9-6. A variable speed motor control with 3 transistors.

Before troubleshooting the electronic speed control circuits, isolate the problem to either the motor or its electronic control circuit. The motor may not rotate, run slow or run fast, or it may operate intermittently. First, check for applied voltage to the motor. If voltage is not reaching the motor circuit, suspect a defective cartridge switch, motor switch, or open wiring. In many cases, a defective motor will have burned wiring or charred choke coils found in series with the supply voltage. Check the continuity of the motor winding with an ohmmeter.

Intermittent motor speeds can be caused by a defective transistor, electrolytic capacitors, or the motor itself. Don't overlook a poor wiping contact on the speed control. The motor may change speeds when the track-shift switch is changed to another channel. If this is the case, remove the motor belt. Slow the motor down by grasping the motor pulley between thumb and fingers. If the motor speed returns to normal the mechanical governor switch is defective inside the motor, and the motor itself must be replaced.

Some motors will run at speeds as low as 300 rpm; others may run up to 3700 rpm. The average motor speed is between 2000 and 3000 rpm. When a fast-forward motor switch is used, the motor may run up to 7200 rpm at 12.6V! With the various motor speeds found in the auto tape player, you should always replace the defective motor with the exact part number.

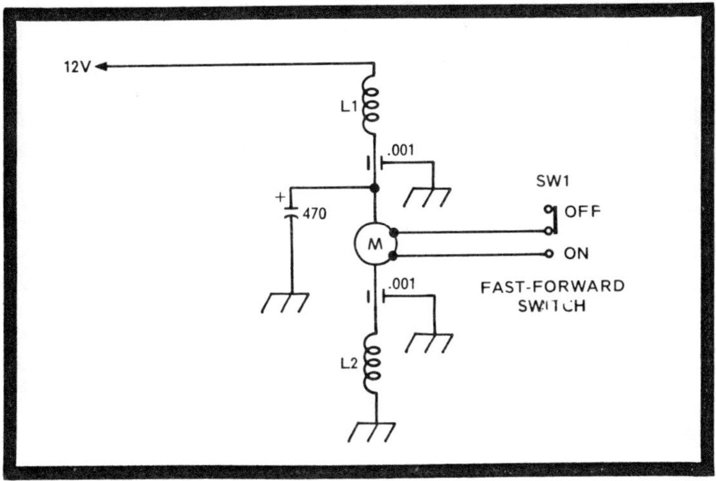

Fig. 9-7. Stereo 8 fast-forward motor circuit.

In *fast forward* (in either the auto cassette or stereo player), higher voltage is applied to the motor windings. In Fig. 9-7 part of the motor winding is shorted with SW1. The transistor is bypassed by SW1 in Fig. 9-8. Instead of 5.5V being routed to the motor winding, the 12V supply voltage is switched to the motor circuit. A shorted or leaky transistor may increase the motor speed in this type of circuit. In Fig. 9-9, the normal motor speed is 3700 rpm. Remove Q2 from the circuit and check it on a transistor tester. While Q2 is out of the circuit, the other two transistors may be checked in the

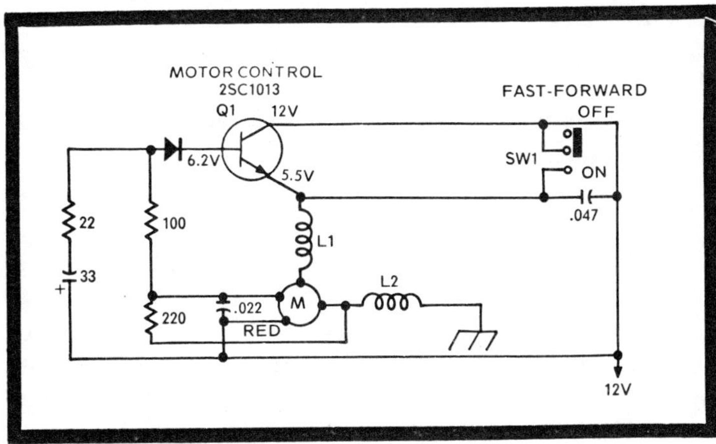

Fig. 9-8. Stereo 8 motor control circuit with fast-forward SW1.

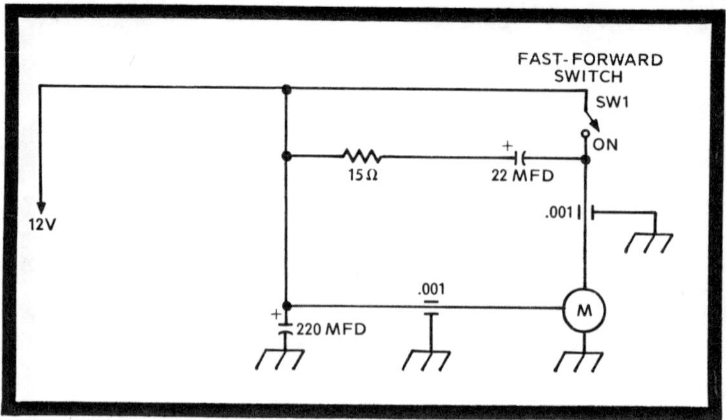

Fig. 9-9. When SW1 is turned on motor speeds up to 7200 rpm.

circuit for open or leaky conditions. Next, check for a shorted or open diode. It's always best to remove one end of the diode from the circuit for accurate resistance measurements.

EXCESSIVE WOW

Excessive wow or abnormal speed is generally traceable to mechanical problems. The flywheel or capstan may be dry. A good cleanup of capstan–flywheel assembly may solve most wow problems. Check the motor pulley or belt for oily spots, since a loose or dirty belt can cause wow. Check for a defective pressure spring or loose pulleys in the cartridge holding assembly.

A defective motor may also cause wow problems. Check the motor for dry or excessively worn bearings. A poorly soldered motor connection has caused many a problem involving intermittent speed. If the motor speed is regulated with a transistorized circuit, check for proper speed regulation.

WON'T CHANGE TRACKS

Most stereo track-shift problems are related to a defective solenoid, open manual switch, and defective ratchet assembly (Fig. 9-10). First, check to see if the solenoid engages by operating the track-shift program switch. Through the front tape opening you can engage the cartridge switch with a pencil or short out the tongs of automatic program switch with a screwdriver blade. If the solenoid does not operate, either power is not reaching the solenoid or the solenoid itself is defective.

An overheated solenoid coil will appear burned and sometimes the insulation will crumble between your fingers. Check to see if the ratchet plunger will slide in and out of the solenoid. A sticky plunger will produce erratic track switching. If the voltage pulls down and the pilot lights dim when the manual channel button is pushed, suspect a shorted suppressor diode across the solenoid winding (Fig. 9-11). Remove one end of the diode and check with ohmmeter.

One of the most unusual track-shift problems occurs when the unit is turned on and the solenoid engages. The track-shift solenoid should only be engaged when the manual or automatic program switch is energized.

In ac models the solenoid action is quite loud compared to the auto tape units. With the covers removed you can see if the solenoid plunger is pulled into the solenoid coil.

A quick way to check out a possible shorted switch or wiring is to apply the ohmmeter leads across the tongs of the

Fig. 9-10. A defective ratchet assembly may produce track-shift problems.

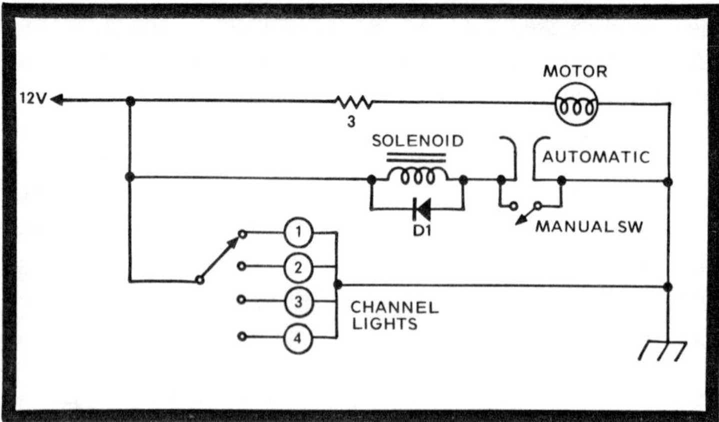

Fig. 9-11. A shorted suppressor diode across solenoid winding will blow fuses when automatic switch is engaged by tape.

automatic program switch. Keep the ohmmeter on the lowest resistance scale. A shorted reading may indicate a defective automatic or manual channel switch. It's possible to have a shorted or defective switch wiring. This reading should be open as both switches are wired in parallel to the solenoid circuit.

Now remove one of the solenoid coil leads going to the automatic switch assembly. The solenoid plunger should pop out. If not, the manual switch insulation has broken down. Sometimes one bent tong will short against another; this allows the solenoid to be energized. Don't overlook a piece of metal filing that has lodged between the two tongs. A piece of foil from a gum wrapper, or a short length of the "switching" foil from a cartridge will cause this.

You may disconnect one switch at a time to find which is defective. Use the ohmmeter for continuity checks. A broken or intermittent manual switch assembly should be replaced. Most automatic change switches are interchangeable, so substitution is no problem in most stereo tape players. Be sure the automatic switch assembly is mounted in the same position and meets the tape.

When the tape player has intermittent or erratic track-shift action, suspect a defective automatic switch or solenoid ratchet assembly. Check the automatic switch assembly for excessive oxide. This switch should be cleaned when cleaning the tape head assembly. Operate the solenoid plunger by hand and see if the ratchet arm moves the tape head. Sometimes these ratchet arms may become bent out of line and will not engage each time.

Excessive Crosstalk

Improper position or angle of the tape head may cause insufficient volume, poor tone quality, and crosstalk. The tape head may get out of position with a jammed cartridge or improper operation of the head lifting ratchet. Foreign material lodged into the cartridge opening will also produce excessive crosstalk, throwing the tape head out of alignment.

Before making any adjustments, inspect the tape head for correct operation. Notice if the tape head is moving up and down the tape area. When the tape head just sits in one position and does not move, either the ratchet assembly is not lifting the head or adjustment is way off. Check for foreign material around the ratchet lifting assembly. (A defective cartridge can cause excessive crosstalk too, so try a known-good one first.)

Generally, excessive crosstalk is caused by improper height adjustment. If the tape head is moving properly you are ready to make height adjustments. Use a head height test tape (500 Hz) on tracks 2 and 6. A simple height and azimuth adjustment schematic is shown in Fig. 9-12. Adjust the height

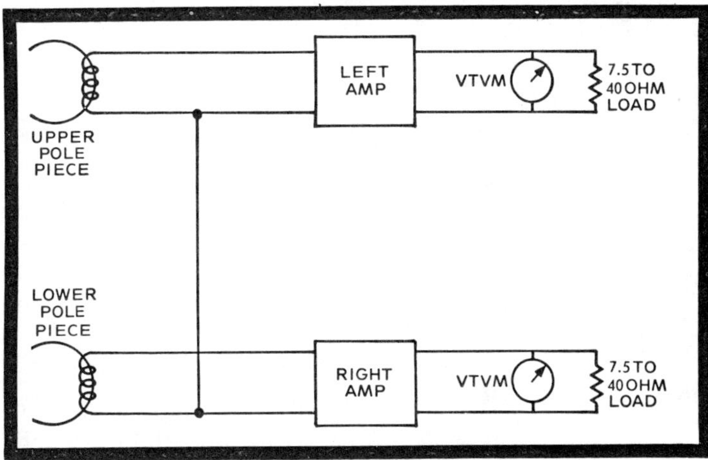

Fig. 9-12. A simple height and azimuth adjustment schematic.

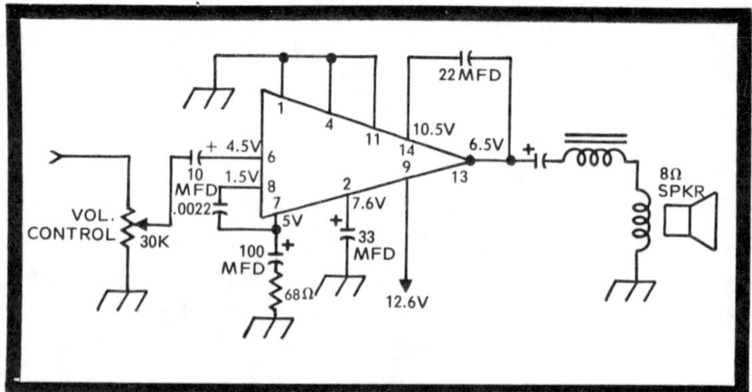

Fig. 9-13. Inject audio signal at terminal 6 to signal trace IC circuit, compare voltages with good channel.

screw for maximum output as measured on a VTVM connected to head terminals. Double-check all channels. Rotate the balance control from left to right for overlapping of tracks. Recheck program numbers to correspond with the correct music on the stereo cartridge.

DEAD LEFT CHANNEL

When only one channel is dead or has a rushing noise, check to see what section the trouble might be in. Inject an audio signal at the volume control. If the signal is good from here to the speaker proceed toward the preamp stages. The stereo tape player may have a stereo FM radio, and you can double-check to see if the problem is in the audio stages by turning the multiplex radio on, and tuning to a stereo station. When either the stereo tape or radio will not play on one channel, suspect the audio section.

Inject the audio signal where the head lead ties to the PC board. Sometimes the preamp stages are found on a separate PC board close to the tape head assembly. If the signal is good here, inject the signal at the tape head terminals. It's fairly common to find a shorted or open shielded connecting cable; try flexing the head cable. The preamp circuit may be an IC or direct-coupled transistor circuit.

The IC can be easily signal-traced with an audio or noise generator. If the IC stage is conventional (Fig. 9-13), inject the signal at terminal 8 on the IC chip (bypass). If this doesn't turn up the culprit, go to the terminal 1 of the IC (or its equivalent pin number on a similar chip). If there is signal at terminal 8 and none at terminal 1, suspect a defective IC. Double-check the IC with voltage measurements.

You may compare these voltages with readings taken in the IC representing the normally functioning channel. If the voltages are way off, replace the IC. With the IC removed from the PC board, make a rough check of coupling and bypass electrolytic capacitors. Check for burned bias resistors or a change in their resistance. Measure the supply voltage for possible defective decoupling resistors.

In case the audio signal is good at pin 1 of the IC, the problem has to be an open electrolytic coupling capacitor or defective tape head. Check the resistance of the tape head on the bad-channel side and compare it with the good-channel side. If in doubt, you can switch leads from right to left terminals on the tape head. A loud hum or audio signal should be heard on either side of the electrolytic coupling capacitor.

WEAK AND DISTORTED CHANNEL

Weak sound produced in just one channel can be isolated quickly by checking the bad channel against the good channel on a stage-by-stage basis. Weak stages can be located with an audio signal generator or audio signal tracer.

Inject an audio signal at the volume control and isolate the weak signal in either the output or preamp stages. (If the signal comes through undistorted, the problem is in the preamp; if not, the output stage is at fault.) Weak signals are caused by leaky or shorted transistors, a change of resistance, and open or leaky electrolytic capacitors. Bad transistors and electrolytic capacitors produce most audio problems in preamp circuits.

If the signal is good at the volume control (center lug, gain at maximum) go to the collector terminal of the second preamp transistor. Then go to base of the same transistor. Keep proceeding through the circuit with the same method until the signal problem manifests itself. If in doubt about the signal at a given point, compare it with the good channel. Always keep the volume of the audio signal generator as low as possible. The audio signal should increase as you go toward the front end of the audio section.

When you have located the defective stage, check the suspect transistor in the circuit. Double-check with voltage measurements. A voltage test of the collector terminal will determine if the collector load resistor has increased in resistance. High collector voltage will indicate an open transistor. A voltage check between base and emitter will indicate the presence of forward bias (if present, the transistor is normal). In-circuit transistor tests will indicate if the transistor is open or leaky. If the leakage test reading is quite high, remove the collector lead and make another test.

To check for a defective coupling or bypass capacitor, bridge a new one across it. If in doubt, remove one end of the capacitor and check with the ohmmeter on the R × 10K scale. A good electrolytic capacitor will charge, sending the pointer up the scale, and then discharge gradually. An open or dried out capacitor will barely charge. Check it against a new capacitor with the same value. A shorted or leaky electrolytic capacitor will show a reading indicating leakage on the ohmmeter.

To check out collector load or emitter bias resistors, remove one end from the circuit. If not, you will get an erroneous reading on the ohmmeter. Remember, the resistance through a transistor is quite low and when paralleled with the suspect resistor, it results in a lower resistance reading. A visual inspection sometimes will reveal a burned or cracked bias resistor. In this case, replace the corresponding transistor.

A dirty or packed tape head can produce weak and distorted conditions. Sometimes this condition may indicate weak signal from both or only one channel. Improper height adjustment of the head will cause weak conditions on either channel. Incorrect azimuth adjustment produces poor high-frequency tone quality.

Weak conditions found in the driver and output stages may be caused by faulty transistors and capacitance and resistance changes. An open driver transistor may produce severely weak signal problems. When one push-pull output transistor opens, distortion always results—particularly on audio peaks. An open or loss of capacitance in an electrolytic capacitor connected between output circuit and speaker results in no or weak conditions, hum, or dc voltage across the speaker terminals.

Distortion problems in the audio stages are generally caused in the audio output stages. (See typical schematic, Fig. 9-14.) Isolate the distorted condition to the output stages by audio signal injection. When isolated to the audio output stages, check the suspect transistors. Check the circuit for burned bias resistors; you should replace both audio output transistors when they are found. Check both bias and emitter resistors for correct resistance. A faulty thermistor can also produce distortion in the output circuits.

If distortion is isolated to the preamp circuits, suspect leaky transistors, leaky capacitors, and possible changes of resistance. In case the preamp transistors test good, suspect a leaky coupling or bypass electrolytic capacitor. Improper collector load resistors can produce audio distortion. When any transistor is removed from the PC board, check the resistance of the emitter and collector load resistors. Weak-and-distorted conditions can be checked with a scope and audio generator.

POOR HIGH FREQUENCY IN PREAMP STAGES

Poor treble can be caused by a number of areas, from the tape head through the preamp circuits (Fig. 9-15). Excess oxide deposit on the tape head is the single most common trouble source. But when you look at the head, check for a worn head surface just to be safe.

Number 2 in popularity is an improperly recorded cassette or cartridge. If a head was oxide-coated when the tape was recorded, no amount of cleaning can bring back those lost highs. So be sure to start your trouble search by using a high-quality test tape or a tape known to contain plenty of high-frequency information.

Bear in mind, too, that prerecorded cartridges are notoriously deficient in highs. (They're recorded at speeds many times greater than playback, and the high frequencies are boosted beyond the recording capability of the

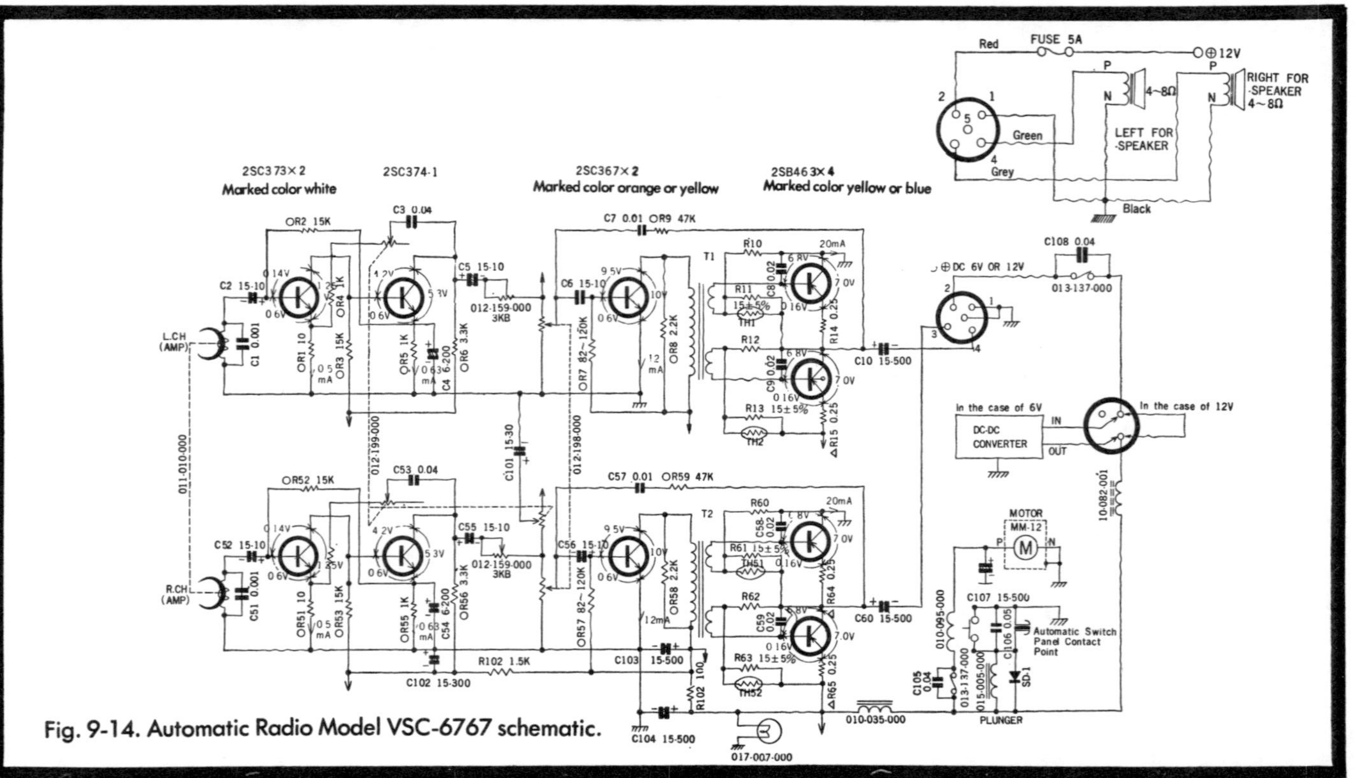

Fig. 9-14. Automatic Radio Model VSC-6767 schematic.

duplicating machinery.) In general, cassettes have a better high-frequency range than cartridges.

If the problem is not in the heads, check the bypass capacitors (if any) shunting the head (Fig. 9-15). If this is of no avail, suspect a problem in the recording-bias circuit. If the bias voltage is set too high, the treble drops off drastically.

When low frequencies are not produced, suspect a defective negative feedback circuit (Fig. 9-16). In the preamp circuits check or replace the feedback capacitor from collector to base of the second transistor. A defective resistor in this same feedback circuit may have changed resistance. Look for a defective feedback capacitor from the output circuit to the base driver circuit of the output stages. Measure the resistance of the feedback capacitors.

MOTORBOATING AND OSCILLATION

Motorboating and oscillation are generally produced in the front end of the amplifier circuits. A defective tape head,

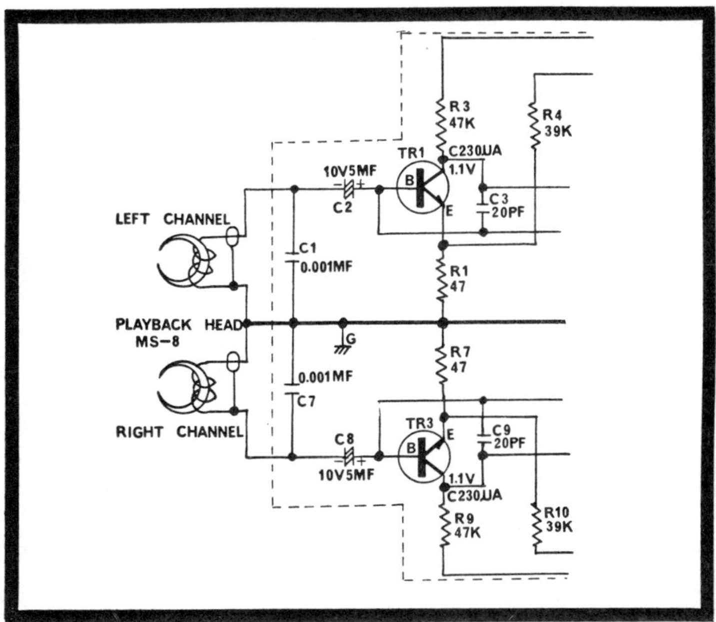

Fig. 9-15. For poor high-frequency treble motor check the heads and their bypass capacitors.

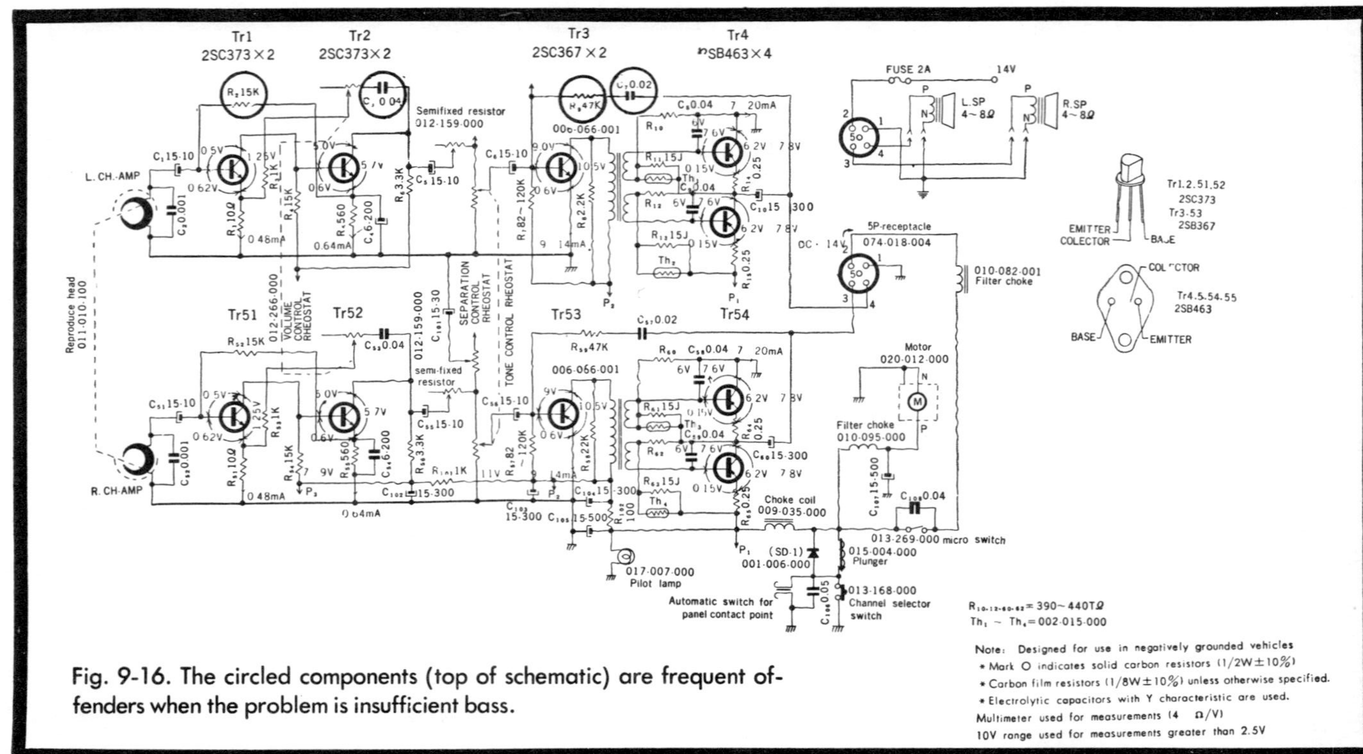

Fig. 9-16. The circled components (top of schematic) are frequent offenders when the problem is insufficient bass.

transistors, resistors, and capacitors can cause either problem (Fig. 9-17). First, note in which channel the oscillations are occurring. Check to see if the volume or balance control has any effect on the oscillating condition. This will help you locate the defective stage.

In models incorporating both radio and tape player, check to see whether oscillation occurs in one position or both. Wiggle the switch assembly and notice if oscillation ceases. A dirty *radio–tape* switch may produce motorboating and oscillation. Clean the switch with tuner spray. Replace the switch assembly if cleaning does not solve the oscillating condition.

Improper tape head wiring or a defective head can produce oscillation. Check for open tape head leads and solder the connections. Look for poor ground connection of the tape head and amplifier. Intermittent oscillations and whistle conditions can be caused by a defective cartridge.

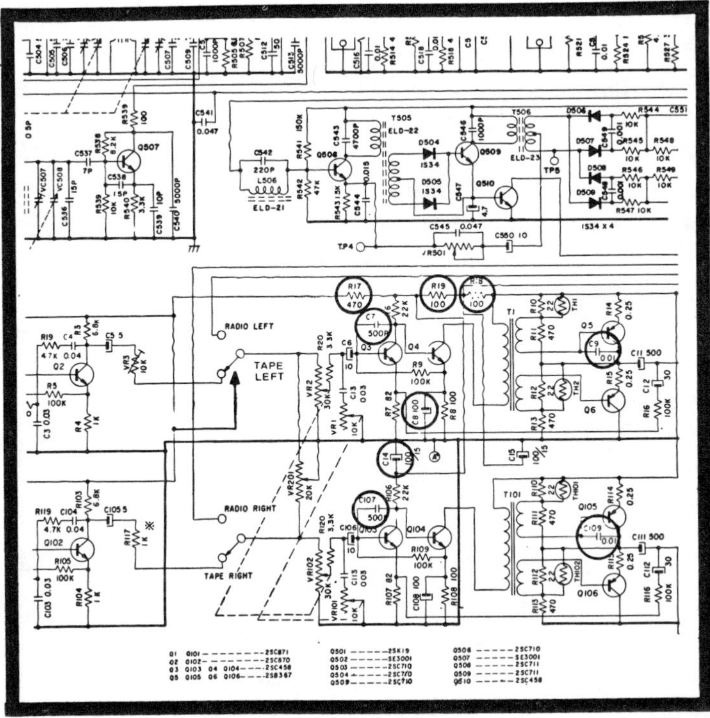

Fig. 9-17. Check the components that are circled for motorboating and oscillations.

Try a new cartridge. Check the suspect cartridge in another tape player.

Off-value decoupling and voltage-dropping resistors in the preamp and driver stages may produce oscillations. Check these resistors for burned conditions or a change of resistance. Open electrolytic capacitors in these same circuits can cause motorboating. Turn off the power and shunt a new capacitor across the suspect.

AUDIO BALANCE PROBLEMS

When channels aren't the same level when the balance is set to mid position, start with a known-balance tape. Suspect a weak audio stage or defective balance control. Most stereo systems will balance around zero on the dial. In some models good balance may be off 1 or 2 numbers on the balance dial. But when the balance control must be turned one-fourth of the way to either channel, you should try to locate the balance problem.

First, try to determine if the balance condition is in the audio amplifier section. For instance, if the FM multiplex radio and stereo 8 tape player balance satisfactorily in a radio/tape/phono combination, then you know the cartridge in the phonograph section is defective. Likewise, if the phonograph and FM multiplex radio balance, the stereo 8 tape head or its associated circuitry is at fault. In case all three sections are off, the problem is in the common audio section.

Poor balance can show up in the circuit before or after the balance control. You can check out the audio amplifier section with an audio signal generator and scope or balance meter. Compare the weak channel with the good one at each transistor stage. Remember, transistors and electrolytic coupling capacitors cause more weak stages than any other components.

Poor balance may be the result of a defective volume control. For instance, if you cannot turn down the volume on the left channel suspect the volume control. The control may be open at this point. Rotate the balance control to the left channel. Now turn the volume control down. If sound is coming through on the left channel, the balance control is normal.

Let's take a look at a typical balance control (Fig. 9-18). When the balance control is turned to the left channel, the full resistance of the balance control is in the circuit. So it could not be the balance controls in this case; you may have a defective volume control.

Either the volume control has an unbalanced taper, open control, or poor wiring. Sometimes you will find a dual volume control, which achieves balance by allowing individual channel gain adjustments. Check the wiring going to the control. When you cannot turn the volume down on one channel suspect an open ground connection to the volume control.

Generally, a defective balance control will let both channels come in on one channel. Rotate the control at the far ends. An erratic or worn balance control will let the other channel pop in when rotated. This is especially true of the sliding type balance controls. Check the wiring terminals for poor soldered connections.

TRACKING DOWN NOISES

Excessive noise in an audio amplifier section can be quite annoying and sometimes loud enough to drown out the music reproduction. Most noisy components are transistors, electrolytic capacitors, and coupling capacitors. Eighty percent of your noise complaints will be traceable to external

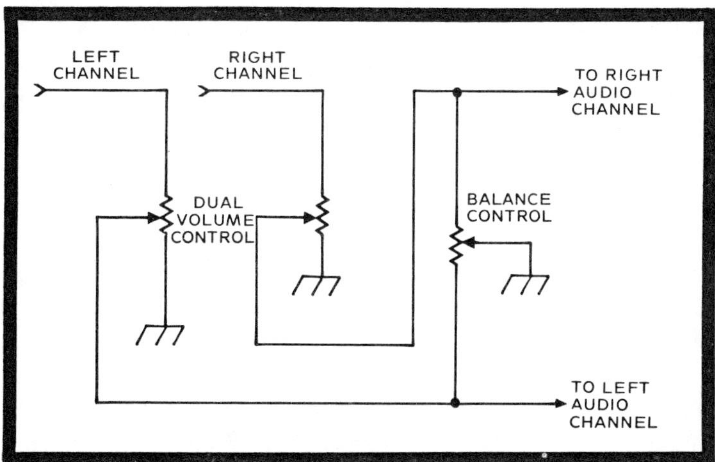

Fig. 9-18. A typical balance control circuit.

interference; of the remaining 20%, the huge majority of ills are caused by audio transistor anomalies. The excessive noise may be in a form of hum, a popping, or a frying sound. Sometimes the noise is intermittent, while on occasion the noise is constantly reproduced.

First, make certain the problem is in the player and not the environment. Then try to isolate what section the noise may originate in. In the stereo tape player, remove the motor belt and disconnect the motor to see if the noise is generated in this section. If not, break the circuit down by turning the volume control down in the noisy channel. Rotate the balance control to see if the noise is produced ahead or back of the balance control. Generally, if the noise is caused by a transistor, it will become amplified and can be lowered or removed with the volume control. If the noisy transistor is ahead of the volume control you can eliminate it by turning the volume down. In case the noisy transistor is a driver or audio output transistor, the volume control will have no effect on it. If noise is found in both channels, the trouble is probably in the power supply.

After isolating the noise in the audio amplifier circuit, try to pinpoint the noisy transistor. When noise is present spray each transistor with *cold spray*. Apply several coats on each transistor before leaving it. If the noise quiets, you have located the noisy component. Then apply heat from the soldering iron to the transistor body or on each terminal. If the noise pops back on, replace the noisy transistor.

After pinpointing the noisy section, you can locate the faulty transistor by removing each suspect transistor from the circuit. Of course, this takes considerable time and effort; but it may be the only way you can locate the noisy transistor in direct-coupled circuits, particularly in view of the fact that the noisy device will probably check out okay on a tester. By removing a suspect transistor you can further isolate the audio section.

Another way you can locate a noisy stage is by using the oscilloscope. Set the vertical gain wide open. Turn the horizontal control to 5 kHz and use the direct probe. Noise generated in dc circuits will make the scope raster pop up and down. See if the noise is found in the dc power supply. If the raster will rise and fall in the power supply section, the trouble could be in any stage that transfers the noise to the power supply. Most noisy problems are rough to find and take up considerable bench time.

Start at the output transistor and work towards the volume control. Go from collector to base of each transistor. Keep going until you receive the largest noisy deflection on the scope or the noise disappears. Generally, the noisy transistor may have noise deflection on both emitter and collector terminals. A noisy coupling or bypass capacitor may have a greater noise deflection in one terminal and lower on the opposite terminal. Remember that noise in a direct-coupled transistor circuit can be produced by either transistor.

Let's check out two noisy conditions produced by a bad transistor and its associated coupling capacitor. In a Sylvania R63-3 solid-state chassis, the right channel was noisy. With the right volume control turned down, the noise was still present. When the balance control was rotated to the left channel the noise disappeared. This meant the noisy component was somewhere in the circuit between the balance and volume control.

In the circuit of Fig. 9-19, the noise may be caused by the IC, electrolytic coupling capacitor, or tone control amplifier Q702. Since transistors are responsible for most of the noisy

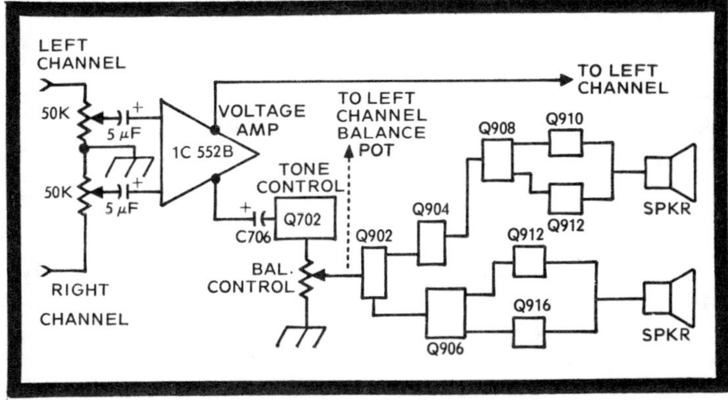

Fig. 9-19. Locating a noisy transistor between volume and balance control.

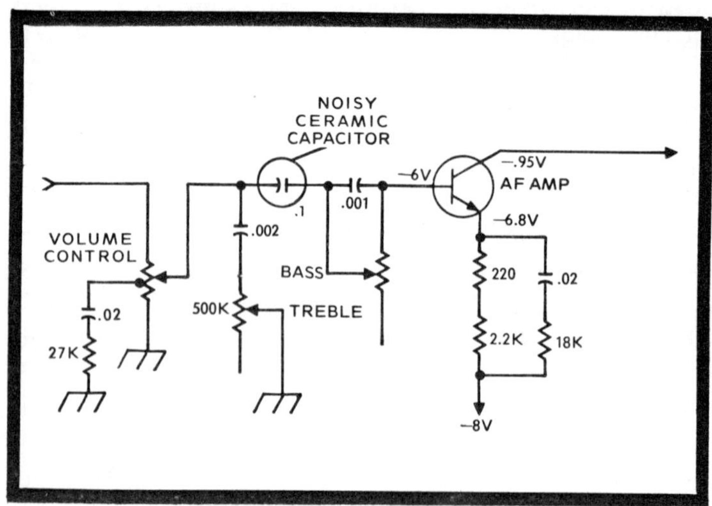

Fig. 9-20. A noisy ceramic coupling capacitor produced noise in audio channel.

conditions, cold spray was applied to Q702. The noise was still present.

The circuit was further isolated by removing the positive terminal of electrolytic C706 from the circuit. When the capacitor lead was removed from the circuit, the noise was still in the right channel. Since all preceding circuits had now been disconnected, this meant Q702 was the culprit. Replacement of Q702 solved the noisy right channel.

In a Morse B6C stereo circuit the left channel was noisy. The noise was still present with the volume control turned way down. This meant the noisy component was between the volume control and speaker (Fig. 9-20). To split the circuit in half, the left driver transistor was removed—and the noise disappeared! This meant the noisy component was still ahead of the driver stage. After replacing the driver transistor, the af amplifier transistor was shorted between base and emitter terminals; the noise was still present, though weak. Replacement of the af transistor did not solve the noisy left channel. The trouble had to be a noisy coupling capacitor. The first step was to replace one arbitrarily. As luck would have it, the first guess was the right one: replacement of the 0.1 μF ceramic coupling capacitor solved the noisy condition.

TROUBLESHOOTING OUTPUT TRANSISTORS WITH OHMMETER

You can check the operation of output transistor pairs with a simple ohmmeter. With push-pull circuits, set the ohmmeter on a low-ohm scale and connect the loudspeakers to the right and left channels. At this time, do not connect power to the tape player. Ground the common ohmmeter lead to the chassis.

Now, start at the speaker terminals on the PC board. You should hear a loud click in the speaker. Compare each ohmmeter test to the other audio channel. Proceed to the metal collector terminal of the output push-pull transistors. Short the emitter and base together with the ohmmeter lead. You should hear a click. Touch the ohmmeter lead to the base terminal and in most amplifiers you will hear a loud hum.

With complementary circuits (a transistor pair of opposing polarity), simply measure the voltage between the common emitter tie point and ground. The potential should be precisely half the total supply voltage.

If removal is required, remove each power transistor individually. You can check the condition of any transistor with the ohmmeter. These checks come in handy when out on a job and without correct test equipment. Of course, it is hard to beat a good transistor beta tester at any time.

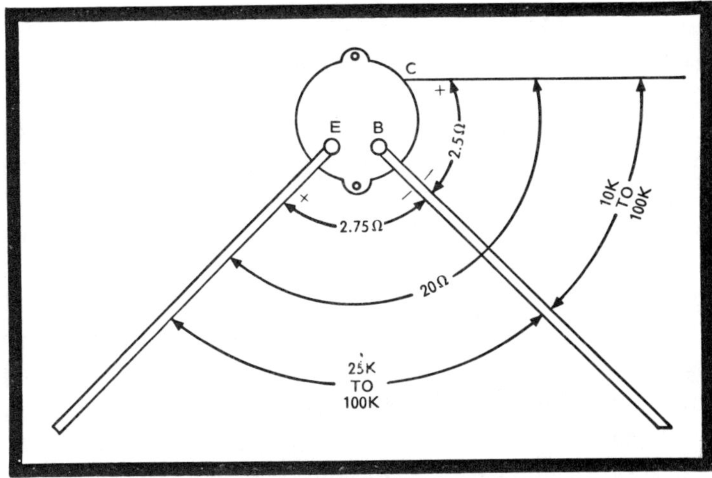

Fig. 9-21. Resistance readings of a typical audio output transistor.

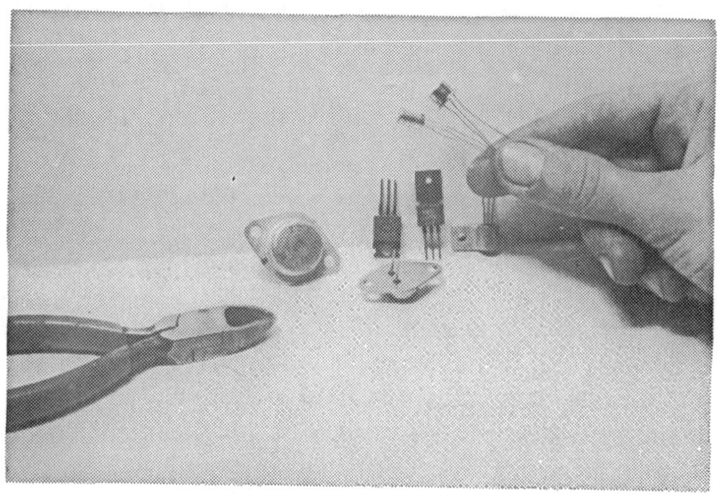

Fig. 9-22. Most power transistors use the chassis as a heatsink. You may find a separate heatsink clipped over the metal container of smaller transistors.

With the positive lead connected to the collector, touch the other lead to the base terminal (Fig. 9-21). You should measure only a few ohms resistance. If straight-through continuity is indicated, the transistor is shorted; with a large resistance, the transistor is open. Reverse the test leads on the same two terminals. You should have a high resistance reading (10K–100K). Place the positive lead to the emitter terminal. The resistance reading to the base terminal should be low. Measure the resistance between emitter and collector (should be around 20 ohms). Reversing the test leads should result in a very high (25K–100K) resistance between emitter and base. These resistance tests are made with the power transistor removed from chassis.

REPLACING POWER TRANSISTORS

After locating a defective power transistor, make sure the corresponding bias resistors are good. Check the condition of the driver transistors. Double-check the other push-pull output transistor. A shorted or leaky driver or push-pull transistor can quickly damage the replacement. It's best to replace both push-pull output transistors when one is found open or leaky.

There are many types of power output transistors found in the stereo tape players. Most output transistors are sink-mounted on the chassis or have a separate heatsink (Fig. 9-22). In many cases one of the transistors will be mounted directly against the metal chassis. The other push-pull output transistor may be isolated with a thin piece of mica insulation (Fig. 9-23).

The output transistor may operate single-ended, push-pull, or complementary. A single output transistor may be driven with several dc-coupled transistors or an IC. Most output transistors can be removed from the chassis by removing the metal chassis screws. The new flat-type power transistors are soldered into the circuit with one mounting screw fastened to the chassis.

When removing a power transistor, notice if a small insulator is found between transistor and heatsink. Generally, you will find one of the transistors grounded directly to the metal chassis. A small amount of silicone grease should be used between transistor and heatsink for better heat dissipation. In many cases the transistor's

Fig. 9-23. Generally, you will find two of the power transistors insulated from the chassis.

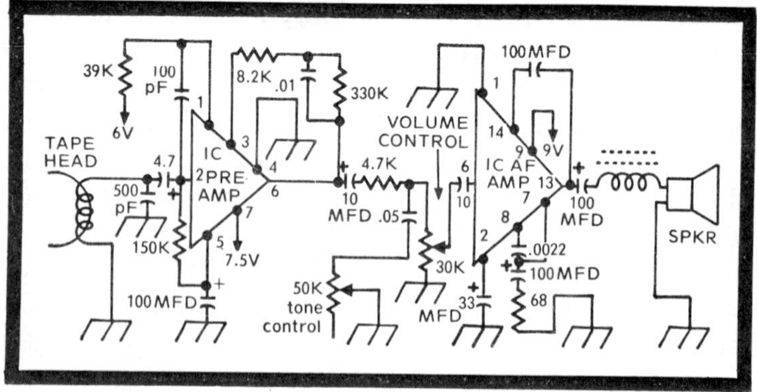

Fig. 9-24. Entire audio circuit may consist of two ICs.

collector is connected directly to the metal case, which serves to dissipate much of the internally generated heat.

After the output transistor has been tested, clean off the old grease and insert a piece of mica insulation. Sometimes dirt or metal filings will cling to the old grease and may short out the new replacement. In most cases a piece of insulation material comes with a new power transistor. Place silicone grease on *both sides* of the insulator.

You can tell if the power transistor is insulated from the heatsink when an insulated washer is found under the mounting screw. Sometimes the insulated transistor will have a plastic screw holding the transistor to the heatsink. Don't make the mistake of replacing this with a metal screw, which would certainly destroy the semiconductor almost immediately.

REMOVING AND REPLACING THE IC

ICs in stereo 8 tape players may be in the preamp, driver, or output stages. In later models, the entire audio circuit may consist of only two ICs (Fig. 9-24). Power-output ICs will typically be mounted onto the metal chassis, which serves as a heatsink. IC terminals are generally soldered directly to the PC board.

Extreme care should be exercised when removing and installing the IC. You can quickly remove the excess solder with a Solder-Wik placed under the soldering iron tip. The soldered terminal will be sucked clean and the IC can be pulled from the PC board. Use a sharp tool or a knife blade under the IC to remove from the etched wiring.

Check the mounting position and tab indication of the IC. Be sure all IC terminals are straight and not bent out of line. The IC may be mounted with the help of a pair of long-nose pliers or tongs. If placed straight with the PC board, the IC will fall into place. You may want to start one whole side and then push the other side of the IC into the corresponding holes. Solder the small terminals with a low-wattage iron. Do not leave the iron too long upon any one terminal or you might damage the new IC replacement. Now scrape off the excess rosin around the IC terminals. You may want to use a small wire brush to clean up the soldered terminals.

SERVICING CARTRIDGE EJECT CIRCUITS

Cartridge ejection circuits may be mechanical or electronic. Electronic systems are usually found in the more expensive players. In the electronic systems, the cartridge will eject when the machine is turned off or if the *eject* button is pushed. In the mechanical ejection system, the front button must be pushed to eject the cartridge. The cartridge should be ejected to relieve the pressure of the pinch roller and tape against the capstan drive. Never manually pull the cartridge out of a player that has automatic ejection as it is possible to damage the eject mechanism.

Most manual ejection systems are the epitome of simplicity. The button itself is mechanically coupled to a pair of simple leaf contacts that resemble the contacts of a relay. When the *eject* button is pressed, one of the contacts, a spring return type, is pushed against the other. Generally, this grounds one side of the ejection solenoid (the other side of which is supplied with a continuous +12V), causing the solenoid to pull in, thus ejecting the cassette or cartridge. The circuit for this simple system appears in Fig. 9-25.

Troubles with mechanical ejection mechanism are generally related to the contacts. Either the button won't touch the leaf contact (as when it's bent out of position), or the contacts remain touching at all times. In either case, troubleshooting is quite simple. Just short the contacts with a screwdriver to see if the solenoid works. If it works, then look

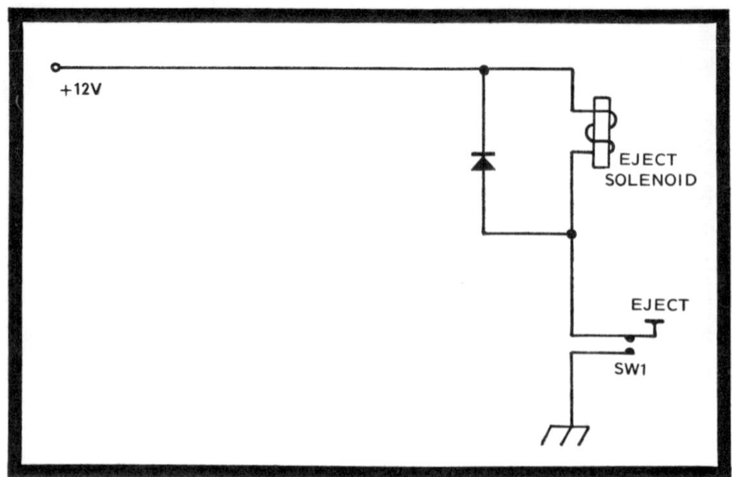

Fig. 9-25. Simple solenoid-controlled ejection mechanism.

for a mechanical problem with the button itself (sometimes foreign matter gets wedged behind the button, preventing it from contacting the leaf switch). If it does not, check the voltage at the ejection solenoid.

There is still another type of manual ejection system that relies on nothing other than the pressure applied to the button. This arrangement involves a linkage that pushes the cassette or cartridge out forcefully when it is pressed. Here again, troubleshooting is not much more than applied brainwork.

Most problems found with ejection mechanisms are defective switches or contacts. But solenoids do go bad, and this is a possibility. When the eject switch is pushed and the main fuse blows, suspect a shorted diode across the solenoid coil.

In Fig. 9-26 the eject solenoid is energized when the cartridge is inserted and holds it in place. When you push the eject button or turn the power off, the holding circuit is broken and the cartridge is released. If the cartridge will not stay in with this type of circuit, suspect a burned solenoid, poor switch contacts, or bent levers of the cartridge mechanism.

Most stereo electronic ejection systems use the solenoid holding circuitry. The solenoid in the collector circuit of the

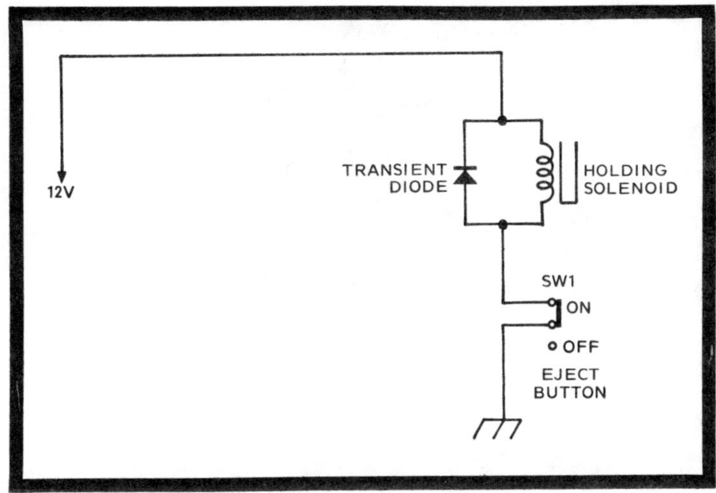

Fig. 9-26. Hold-in type cartridge eject circuit.

power transistor (Fig. 9-27), is energized when the cartridge is inserted. Q2 is biased to pull heavy current through the solenoid winding. When the bias is lowered or removed (as with the eject switch), Q2 pulls less current, thus releasing the solenoid eject lever. The moving solenoid core forces the cartridge outward.

When the cartridge in this circuit will not hold and you've examined the contacts, check the solenoid. You can check the winding with an ohmmeter. If you find the solenoid burned or charred, suspect a shorted or leaky power transistor.

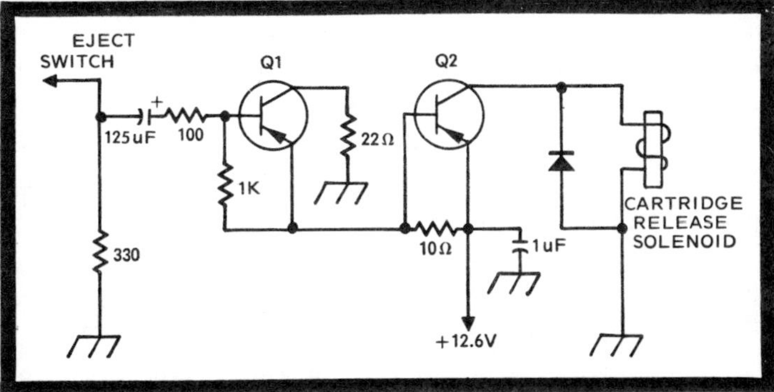

Fig. 9-27. The eject solenoid is located in the collector circuit of the power transistor.

Fig. 9-28. Here is one type of a power supply that any auto stereo unit can be slipped into at home.

Remove the transistor and check it away from the circuit. (You may check Q1 while it is in the circuit.) When a leaky or shorted transistor is found, check the bias resistors for burned marks or a change in resistance.

Another method to check out the solenoid is to place a screwdriver blade near the metal core. The blade should vibrate and pull against the metal core while the solenoid is energized. Double-check with a voltage measurement across the solenoid winding.

As noted, the mechanical cartridge ejection assembly consists of levers and perhaps springs. When the cartridge is inserted, spring tension locks the cartridge in position. The cartridge may be ejected by pushing a button which relieves the spring tension and kicks the cartridge out. Check for weak or bent springs and levers when the mechanical ejection assembly will not hold or release the cartridge. Check the ejector assembly for dry or worn areas. A thin coat of grease may solve the problem.

AC POWER SUPPLIES

A number of auto stereo manufacturers are now marketing a special ac power supply which houses the auto tape player (Fig. 9-28). The player can be removed from the car and plugged into the unit. The component consists of an ac power supply and speaker connections. This allows the auto stereo tape player to be used in the home.

When the auto stereo tape player is inserted into the table unit proper power and speaker connections are made automatically. The ac power supply is a simple full-wave circuit with a very large electrolytic filter capacitor (Fig. 9-29). A protection fuse is included in the center leg of the transformer. Most problems found in these units are shorted diodes and defective filter capacitors. If the centertap lead breaks, however, the supply won't operate without a circuit modification.

DEFECTIVE POWER CONNECTIONS

In the auto tape player you will find a safety fuse between the 12.6V battery and player. These players may pull from 0.4 to 5.00A of current. Fuses may range from 2 to 9 amps, depending on features.

Generally, the fuse lead and container are connected to a male plug that fits into the back of the tape player. Both speaker wires are found in the same plug. To service all types of auto tape players, you should have several different plug-in harnesses. Figure 9-29 shows the cable terminations for most of the currently popular cartridge and cassette players. With these hookups you can easily connect and disconnect to practically any auto tape player that might be brought in for service.

STEREO TAPE PLAYER HINTS

After those small screws have been removed, place them in a small plastic butter container or the plastic top of a hair-spray can. A cardboard box doesn't work out too well, because the screws tend to lodge under the panels. Another good idea is to purchase a box of assorted replacement screws. This can save you a lot of bench time when a screw or two gets lost anyway.

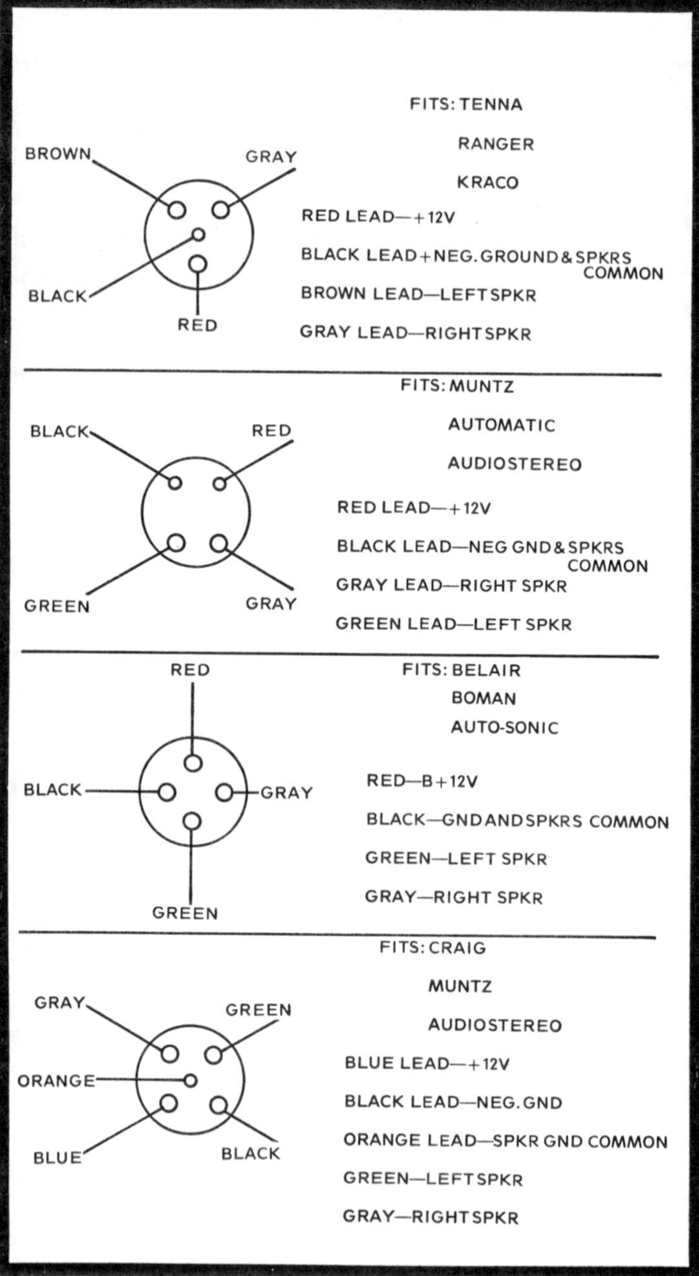

Fig. 9-29. Power supply connections will fit most plug-in auto stereo tape players.

In most imported stereo units you will find various colored screws. The red-painted screws are typically those that hold the tape player to the main chassis. Just remove these screws and the player can be lifted out of the chassis.

Select an assortment of imported Japanese pilot-light bulbs for channel indicator lights. Most are bayonet types and can be held in place with glue or metal clips. Also, silicone rubber cement works nicely here. These bulbs should have long pigtail hookup leads (they may be cut as necessary for correct length).

When replacing a component with a lot of leads, such as power transformer or motor, leave about one inch of the soldered lead. Now you can replace and match the exact component and wire up the correct leads. Always cut off the 1 in. piece after lead is soldered in place. It is wise to make a note of the color-coded leads before removing any component.

On some auto cartridge tape players, the top and bottom covers are held in position with small screws. Be careful not to scratch or mar the cabinet with a screwdriver. Lay the tape player on a drop cloth or pad to protect the metal covers. When replacing the top be sure to place the cartridge lid spring under the top cover.

Be sure and remove your test or blank cartridge from the tape player after repair. It is best not to receive any stereo cartridges with the tape player unless they are to be checked for operation. You can paint or mark your own test tapes with your favorite color so they can be removed.

Intermittent operation of the channel lights can be caused by a dirty channel switch. Clean the switch first, before replacing the small channel light. Some of these bulbs are glued in place and are difficult to remove.

Always keep a load on those output transistors. Plug the speakers in on both channels or clip an 8-ohm resistor (10W) across the speaker terminals. Make it a habit to hook up the speakers before the tape player is turned on.

When an auto tape player comes in with damaged areas or excessive scratches you can touch up the case with spray paint. These units are generally black and one coat does wonders for the cabinet appearance. Mask off the front chrome panel with masking tape and paper. One can of spray paint may last all year and may cost less than a dollar bill.

Use window cleaner spray to clean up the metal case and front chrome piece. Spray the cleaner over knobs and in hard-to-get-at places. Use a small brush for those crevices and wipe off excess cleaner with a soft cloth.

TROUBLESHOOTING CHART FOR AUTO CARTRIDGE PLAYERS

Trouble	Cause
No tape motion; pilot bulb lights	Setscrew loose on motor pulley
	Motor frozen or burned out
	Drive belt off
	Capstan drive frozen
	Pinch roller seized up
	Broken motor connections on PC board
Excessive wow or flutter	Defective tape cartridge
	Loose or defective drive belt
	Oily drive belt
	Improper lubrication
	Defective or oily capstan
	Defective motor
	Frozen pinch roller
	Flat spot on pinch roller
	Defective spring and rubber assembly
	Dry capstan bearing
	Check electronic speed circuit
Excessive noise	Defective motor
	Dragging capstan—flywheel
	Noisy motor bearings
	Capstan frozen or seized up
	Dry capstan bearing
	Defective cartridge
	Ignition (auto) noise
	Demagnetize playback head

No or low audio output	Defective cartridge Speaker problem (wiring, voice coil) Defective amplifier components Tape head out of adjustment Excessive oxide deposits on tape head Low battery voltage (only on battery units) Use audio signal generator to locate weak stage
Excessive crosstalk	Defective cartridge Improper head height adjustment Defective cam assembly Foreign material jammed in cartridge opening
Low treble	Dirty head Worn head surface Head alignment Amplifier component
Will not change channels (automatically)	Shorted or open solenoid circuit Shorted silicon suppressor diode Shorted or open selector switch Defective pawl Open or shorted solenoid coil Defective cam mechanism Head index cam frozen
Will not change channels (manually)	Defective manual switch Dirty switch contacts Broken wires to manual switch Indicator arm binding
Blows fuses when changing channels	Check diode across solenoid winding (suppressor) Shorted transistor Shorted supply lead
Excessive distortion	Defective tape Dirty head

Unbalanced audio output	Defective speaker Improper bias Defective amplifier section Improper recording volume Low dc supply voltage Speaker wire too fine in gage Excessive dc supply voltage Poor speaker connections Defective balance control Bad tape head Defective speaker Incorrect height adjustment Check for weak auto stage
Tape runs too slow	Belt slipping Hub binding Oil on belt or capstan drive Bad cassette or cartridge Worn or dry capstan bearing Defective motor Motor pulley loose Defective electronic speed circuit
Poor sound caused by power supply	Faulty cartridge switch Faulty output connections (cable plug) Open line choke Open resistors to circuits Leaky filter capacitors
Sound problems in output circuits	Open speaker (voice coil) Open speaker electrolytic coupling capacitor. Open base driving transformer PC board or wiring problems Open or leaky output transistors Open or burned bias resistors
Sound problems in driver stage circuitry	Open or leaky driver transistor Open or leaky base coupling capacitor Improper bias resistance

Symptom	Possible cause
Sound problem in preamp stage	Open or leaky transistors Open balance control Open or leaky coupling capacitors Open or shorted tape head leads Defective tape head Improper bias resistance change
Poor quality or noisy sound	Check deposits on tape head Demagnetize tape head
Poor sound volume in output circuitry	Open or leaky driver or output transistors Open base driver resistors Open or leaky output coupling capacitors (electrolytic) Burned bias resistors
Poor volume in preamp circuits	Open or leaky preamp transistors Defective voltage dropping resistors Defective tape head Dirty tape head Leaky emitter bias capacitor
Poor highs in preamp circuits	Incorrect head azimuth (head angle) Defective tape head Open capacitor across tape head Defective electrolytic coupling capacitor of tape head circuitry
Poor lows in feedback circuitry	Leaky feedback capacitors in preamp and output stages Open feedback capacitor Change of resistance of feedback resistors
Distortion in output and driver	Leaky or open output transistor Shorted or leaky base capacitors Shorted or leaky speaker electrolytic coupling capacitor Open or burned bias resistors

	Open or shorted driver transformer winding
	Open or burned base bias resistor
	Open bias resistor in driver circuitry
Distortion in preamp stages	Leaky or open transistors
	Open or burned collector resistors
	Open or leaky electrolytic base coupling capacitors
Oscillations produced in power supply	Filter capacitors
	Change of value of resistances between power and driver stages.
	Open decoupling electrolytic capacitors
Oscillations found in output stages	Change of value of feddback resistors
	Open or leaky feedback capacitor
	Open resistor across primary winding of driver transistor
Abnormal oscillations in driver	Open bases bypass capacitors
	Bad feedback capacitors
Abnormal oscillations in preamp	Defective tape head leads
	Defective balance on value controls
	Defective or dirty radio—tape switch

Servicing the Portable Cartridge Tape Player

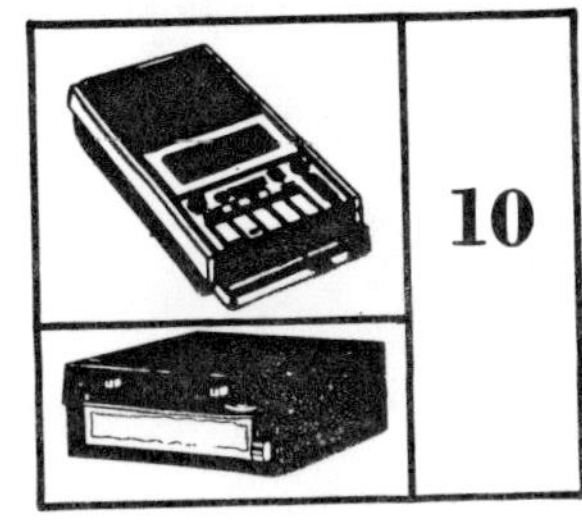

10

The portable 8-track tape player may be operated from batteries, car, boat battery, or ac power. This portable tape player may be monaural or stereo. The monaural tape player typically has one carrying case while the stereo tape player opens into two separate containers so you can extend the speakers for stereo separation.

To operate the portable tape player, disengage the two sections by lifting the top and side clasp. The two speakers can be separated by disengaging the bottom hinges and pulling the unit apart. Remove the speaker connecting cord from the small compartment of the rear and the right speaker section. Plug the speaker connector cord into the connection socket on the left speaker section.

The tape player is turned on by inserting the cartridge into the slot as far as it will go. Remember to insert the cartridge with the label side towards operator. Most tape cartridges are inserted into the side of the left section. The tape player stops automatically when the tape is removed.

You will find the monaural tape player will not change channels automatically. A knob must be rotated to the right to select the desired channel. Numbers appearing on the knobs identify the operating channel. Also, you will find a single channel of audio powering a small PM speaker.

Figure 10-1 shows a typical power supply of the type you'll find integrally included with most "serious" portables.

AMPLIFIER CIRCUITRY

The typical portable amplifier section consists of 5 or 6 transistors in each stereo channel. A conventional transistor amplifier section consists of three stages of amplification, plus a class B power amplifier push-pull output stage. The

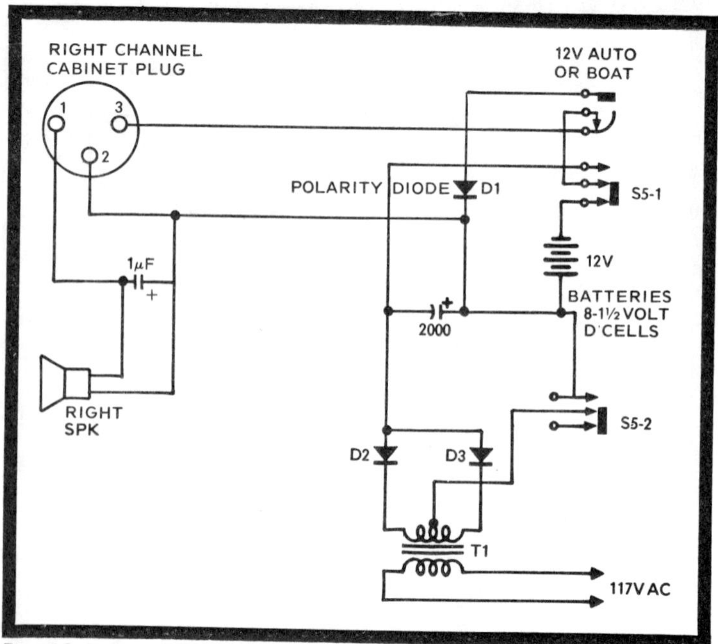

Fig. 10-1. Portable stereo 8 player may operate from car, boat, or household power.

preamp stages are direct-coupled with a driver transistor that is transformer-coupled to the push-pull audio output stage. Some portable stereo circuits use the comp-sym (complementary-symmetry) arrangement shown in Fig. 10-2. The power amplifier uses thermistors to stabilize the thermal characteristics of the device. Negative feedback is employed to reduce audio distortion.

Since power requirements are small (between 1 and 2 watts of output power), the stereo amplifier is mounted on a small PC board (Fig. 10-3). These boards are jammed up against the mechanical section, but they can be removed with relative ease for service operations. The amplifier section can be signal-traced in the same manner as any stereo unit. Solid-state amplifier problems in portable stereo players are the same as any stereo amplifier.

POWER SUPPLY CIRCUITS

The big difference in the power supply circuits of the portable stereo tape player are the built-in provisions for

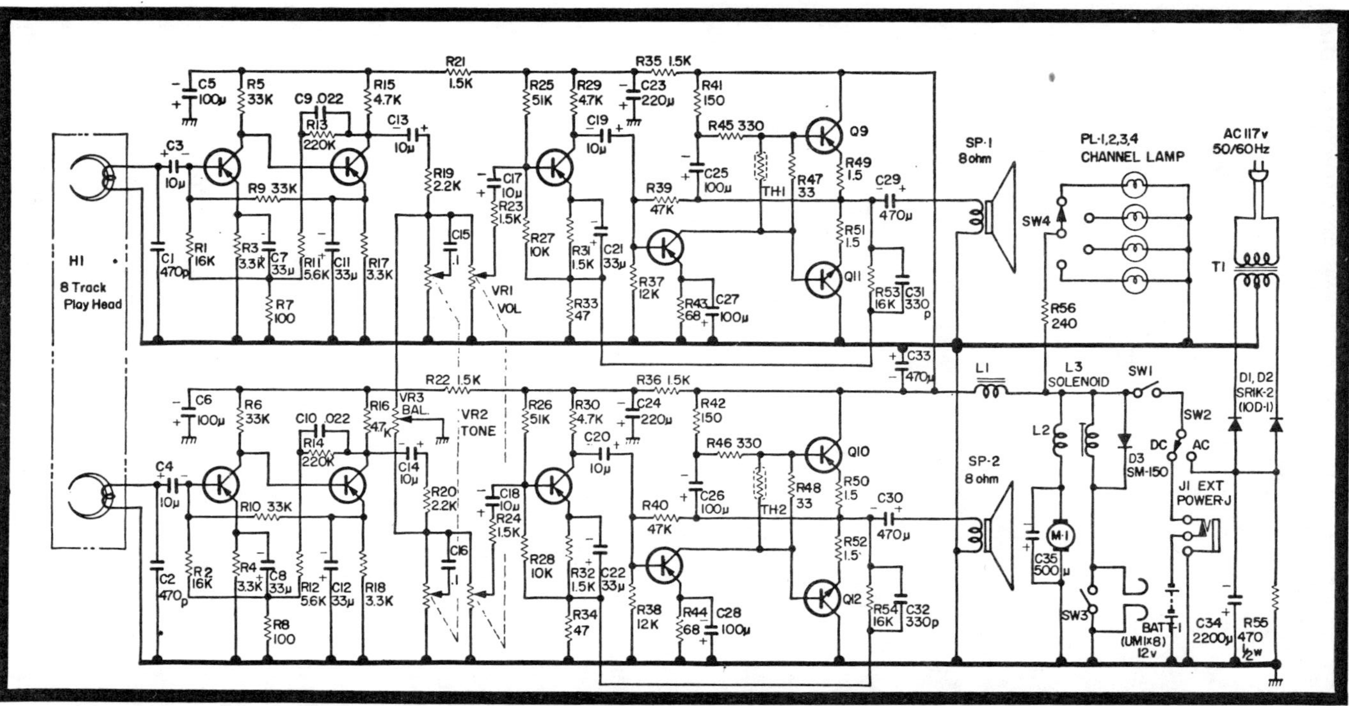

Fig. 10-2. Typical portable tape player with "comp-sym" output. (Courtesy Lloyds.)

Fig. 10-3. The audio amplifier circuits are quite compact; here both stereo circuits are mounted on the same PC board.

ac–dc switching (Fig. 10-4). Generally, eight flashlight batteries provide the dc for 12V operation. The biggest complaint in battery operation is slow speeds, weak output, and inoperative channel track-shift operation. When the player is to be left inoperative for an extended period, the batteries should be removed to protect the player from corrosive damage caused by leaky cells.

If the complaint is that the tape player won't operate on batteries, suspect weak or improperly polarized batteries. If you'll pretend there are no batteries in the set, you'll always connect your own power supply first-thing...and you'll save yourself a lot of time in the long run.

When batteries have been left in the portable tape player for a long time, the cells will leak and corrode the battery contact springs. These metal springs must then be cleaned and scraped for good dc contact. You may find this to be the only trouble with the small portable tape player. When the tape player will not change channels, suspect weak batteries. Batteries may test good but still not operate the track-shift solenoid, which draws a considerable amount of current. A slow motor speed and lower power output may also be caused by weak batteries.

The ac power supply circuits are quite common to all stereo 8 tape players. Either a full-wave or bridge rectifying system is used. In larger units, with a combined radio, bridge-type rectifiers are almost universally found. Notice how the external car and boat battery power is switched into the portable stereo power supply circuitry.

TRACK-SHIFT OPERATION

The portable track-shift solenoid operates like any 8-track tape player. Track indications are identified with

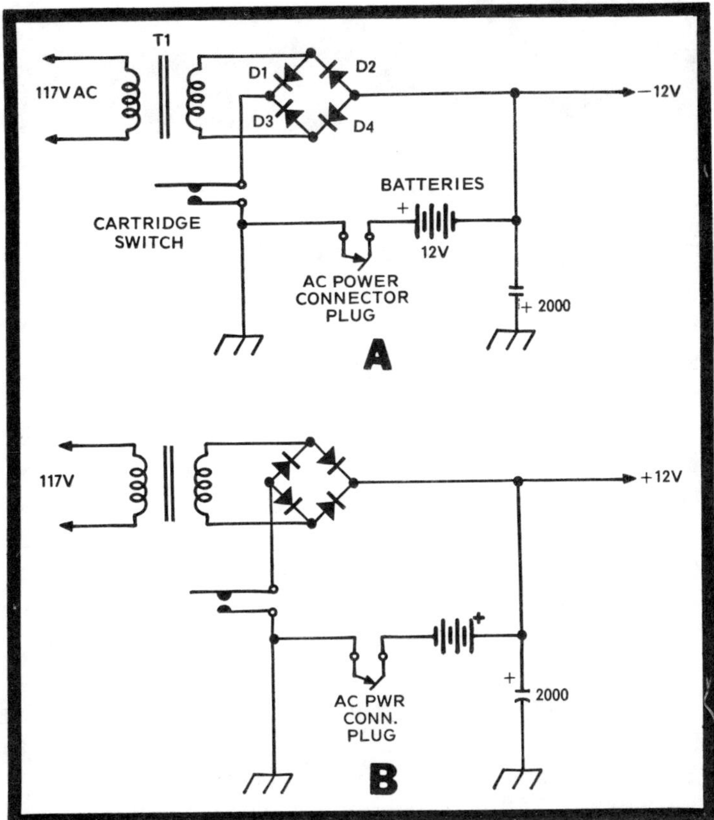

Fig. 10-4. Bridge rectifier circuits with batteries. (A) Positive-ground chassis; (B) negative-ground chassis.

Fig. 10-5. Some portable stereo tape players' track indicators have rubber-toothed belts or gears to turn the track numbers.

pilot lights and geared dials. Some track dials, such as the one shown in Fig. 10-5, are turned with rubber-toothed belts. When the track-shift solenoid operates intermittently or will not energize correctly, suspect weak batteries in battery operation.

Check the ratchet assembly if the track-shift assembly will not raise or lower the tape head. In some models the tape head is raised by a head pin extending up through the deck chassis. A nylon track-shift assembly pushes or lowers the tape head. If excessive tape oxide collects around this metal pin it will not return to the lowest channel. When the tape head is erratic and will not move off program 1, wash out the excessive oxide.

Trigger the solenoid assembly by hand and watch the tape head movement. Since the head pin is located directly beneath the tape head, tape oxide tends to pile up around it. You can also have excessive crosstalk with the same condition.

It's possible to have an open or shorted layer of wire in the solenoid (Fig. 10-6). Very few solenoids found in portable tape players become charred or burned. Either the batteries or rectifier diodes will give way before extensive damage occurs to the solenoid winding. Check for dirty or poor contacts of the indicator switch when channel lights appear intermittent. A broken channel switch may leave one channel on all the time or make two channel lights come on at once.

NO TAPE MOTION

When the tape does not move can you hear the motor run? Is the pilot light on? Now turn up the volume control and listen for a loud hiss.

Open up the cartridge slot and push the cartridge switch with a long pencil or pen. If the motor pulley is turning and the flywheel does not move, check for loose or broken motor belt. If the motor pulley does not rotate, check for an open fuse, defective cartridge switch, or open wiring.

Remove the tape deck assembly from its cabinet to check cartridge switch and wiring. If the drive belt is loose, see if the motor pulley turns freely. Now spin the capstan–flywheel assembly. A frozen or loose capstan–flywheel will throw rubber belts. The motor belt

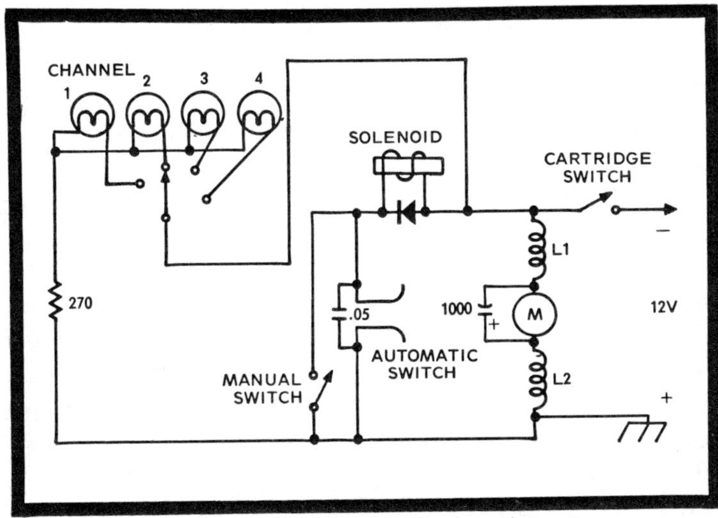

Fig. 10-6. Track lamp and solenoid circuit.

will not stay on if loose or excessively worn. Check the belt for cracked or oil spots.

Push the cartridge switch to see if it is making contact. Check switch continuity with ohmmeter. Sometimes these switches have a broken spring or leaf and may not make contact. Take a dc voltage reading at cartridge switch and motor. Remember these small motors in portable tape players are dc operated. They either operate from batteries or the dc power supply.

When the solenoid is energizing and the channel lights are working you know that dc power is applied to the motor circuit. If the motor does not turn, either the choke coils are open or the motor is defective. Make a continuity check of the motor winding with the VOM.

WILL NOT CHANGE CHANNELS

If the tape player will not change channels in battery operation, replace batteries. Check voltages in both battery and ac operation. If the tape does not change channels automatically, suspect a broken shorting switch. Clean surface contacts of switch.

NOISY CHANNEL

Actually, noise can be produced in any stage of the solid-state amplifier. Noise can be generated from arcing motor brushes and picked up by the sensitive amplifier. A worn capstan or motor bearing may cause a mechanical noise. Try to isolate the stage or component producing the noisy problem.

Rotate the balance control to see if the noise exists in the opposite channel. If so, the noise is common to both channels and may be in the power supply. Measure the dc voltage with a good VOM to see if supply voltage changes in coincidence with the noise. Turn the volume control all the way down. Is the noise still there? If so, look for the problem in the final amplifier. If not, start checking the preamp circuits. (The noise dividing line is at the volume control, as shown in Fig. 10-7. If we turn down the volume control and noise is present, the noisy transistor may be the driver or push-pull output stage.)

Power transistors have a tendency to be noisy, so let's start there. Testing any transistor in or out of the circuit will not show up the noisy problem. If the output transistors are power types and screwed to the heatsink we can remove one at a time until we have located the noisy culprit. In case the output transistors are of the top-hat variety with heatsink, simply short the base and emitter terminals together. You may want to go a step further and remove the base terminal from the PC board. When the noise stops you have found the noisy transistor.

When the volume control is turned down and the noise disappears you can bet it is in the front-end stages. We can quickly isolate the af stage by disconnecting the electrolytic coupling capacitor. Now, if the noise is still present the af transistor is at fault. In case the noise disappeared, remove the first preamp transistor. Now, turn on the amplifier once again and see if the noise has disappeared. Replace the entire preamp stage when a noisy IC is involved.

In vacuum tube audio circuits you can use a bypass or electrolytic capacitor in locating noisy stages. Just clip the capacitor to chassis, ground and touch the grid to each audio tube. When the noise disappears between two points you are close to the defective component. It is best not to attempt this isolation method in solid-state servicing; a charged electrolytic capacitor can damage the small transistors.

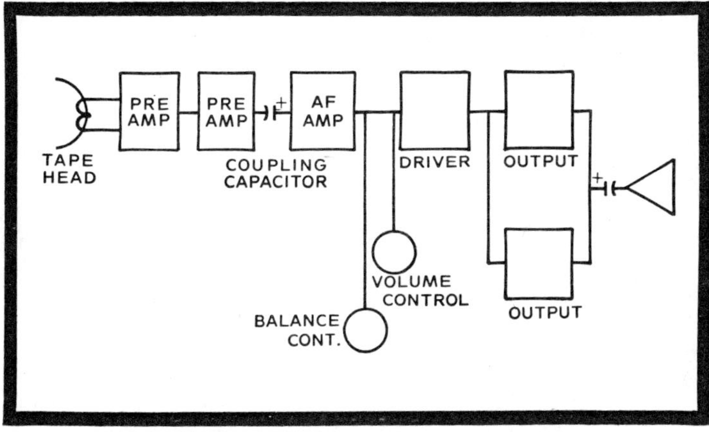

Fig. 10-7. Isolate the noisy right channel.

MOTOR RUNS BACKWARDS

Check to see if the motor runs backwards in both ac and battery operation. If only in battery operation, remove the batteries, and reinstall them with attention to correct polarity. In case the motor runs backwards with external boat or auto operation see if the cigarette lighter cord has been tampered with. It is possible someone tried to repair the male plug end and reversed the two wires. You can check this condition with an ohmmeter.

Another item to watch for is incorrectly resoldered wires from some previous "repair" job. Be particularly watchful when you have reason to believe the owner may have been tampering with the circuit. Look for reversed power-supply diodes, reversed supply leads, and other similar problems.

The portable tape player should not be operated with motor running backwards for any length of time. The tape snarls under this condition are legendary. If only a small amount of tape has been pulled, you may salvage the cartridge by feeding the tape back into cartridge. Use your thumb and fingernail and push down against the tape—pushing excess tape into the cartridge opening. Whether the effort is worth the time is a moot question.

When excess tape spills from the cartridge you can bet the motor is running backwards. Carefully remove the cartridge and insert a blank cartridge. Now take voltage polarity readings of motor and amplifier. Make a quick check. You can't hurt the motor but it's possible to damage transistors with reversed polarity. So be careful or the job can turn out to be a nonprofit repair.

LOW OUTPUT

Low sound output is usually the result of a dirty tape head. This is especially true if low treble is noted in recordings. After cleaning the tape head, check for head alignment. If weak sound is only coming from one speaker, suspect a defective amplifier stage, You can use the other good channel to help track down the trouble.

Most weak audio problems are caused by transistors and coupling capacitors. Generally, most weak-sound problems are combined with excessive distortion. Signal injection is the best method to locate a weak stage. After the weak section is located, inject the same signal (at the same point in the good channel) to verify your "find."

POWER SUPPLY PROBLEMS

Problems found in the power supply of the portable tape player may produce no tape motion or no sound. Check to see if the player operates on batteries. If so, the trouble has to be in the ac power supply section (see typical full-wave circuit of Fig. 10-8). If one of the diodes opens, the output voltage is halved but the unit may still be partially operational. Check both diodes with an ohmmeter for a high front-to-back resistance ratio. (Disconnect one diode end before checking.)

Check for ac voltage at the primary terminals of the power transformer. Now check for low ac voltage on the secondary side. You may have ac voltage on the primary of the power transformer and the winding may be open. Touch the ohmmeter leads to the ac plug and see if continuity is present. If not, check each ac wire lead to the power transformer for a possible break. Now check for complete continuity across the primary winding wire leads. You will find the primary windings of these small power transformers will open up, since they are wound with very fine wire. When

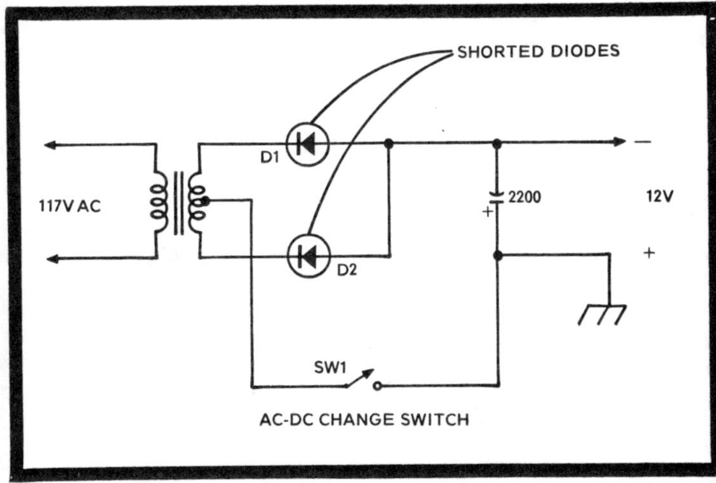

Fig. 10-8. No power supply voltage (shorted silicon diodes).

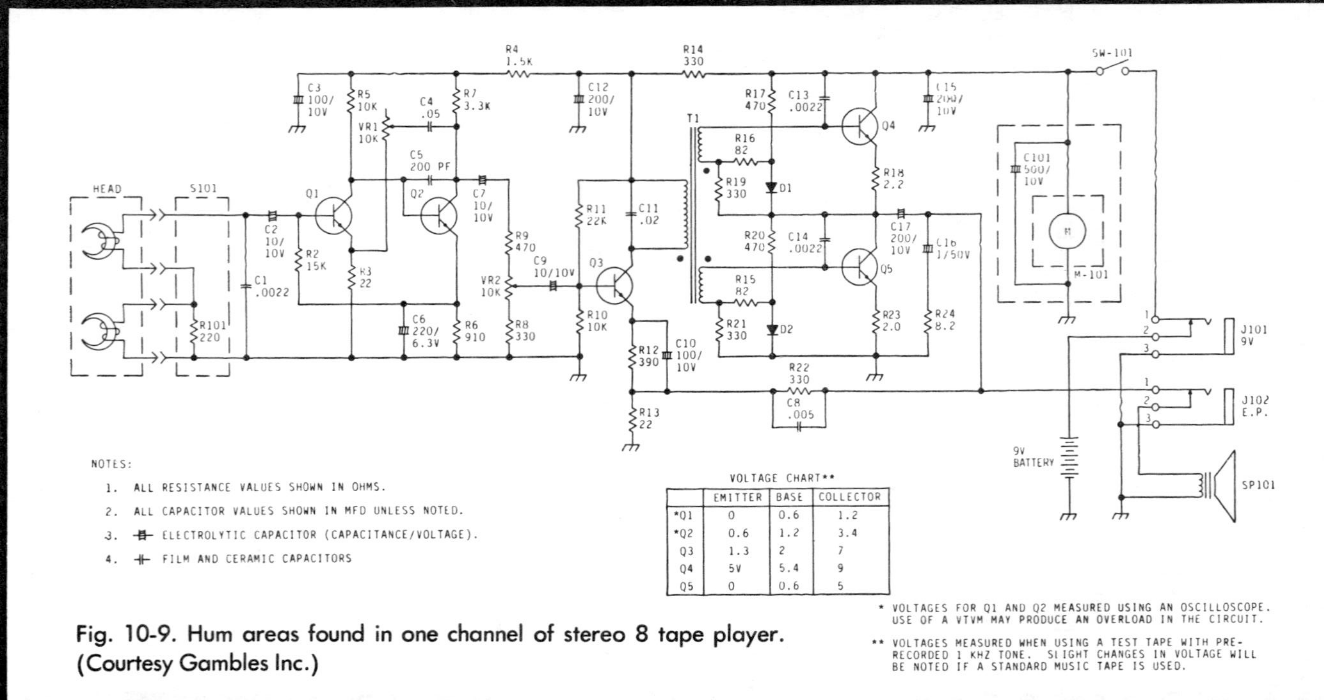

	EMITTER	BASE	COLLECTOR
*Q1	0	0.6	1.2
*Q2	0.6	1.2	3.4
Q3	1.3	2	7
Q4	5V	5.4	9
Q5	0	0.6	5

Fig. 10-9. Hum areas found in one channel of stereo 8 tape player. (Courtesy Gambles Inc.)

a defective transformer is found, *always* replace the silicon diodes. A shorted silicon rectifier will quickly destroy the power transformer.

Excessive hum found in the portable tape player may be caused by leaky or open filter capacitors. Other defects are shown on the schematic of Fig. 10-9. Check to see if one of the large paper filter capacitor leads may be broken off. This may occur in portable tape players and radios. Don't shunt a large filter capacitor across a suspected filter while the player is on. Clip the capacitor into the filter circuit with test leads. Observe correct polarity and insert the tape cartridge. If the hum disappears, you have located the power supply problem.

Install a new filter capacitor with the correct capacitance and working voltage. Never replace the defective filter with a lower working voltage. In case the old capacitor voltage markings are gone and a diagram is not readily available take voltage measurement. Most portable tape players work at a voltage of around 12V. Replace electrolytics with at least a 16V rating.

If the capacitor replacement has a low working voltage it will run very warm. Feel the electrolytic capacitor after a few minutes of operation; if the outside case is warm replace the component. If not, the capacitor may heat up until it explodes. You should replace the defective filter capacitor with the same or greater capacitance. A large capacitor will provide better filtering in any tube or solid-state power supply. Don't tie up a service job when exact filter capacitors are not on hand—use a larger capacitance.

For instance, if the defective filter capacitor is 1000 μF at 16V, replace it with anything up to 2000 μF with a 16, 20, or 25V rating. If a larger capacitor (physical size) is used, tie it down so the leads will not break off. Of course, physical size will play a big part in replacement. Sometimes this small portable tape player may not have adequate room for a larger filter replacement. When you buy replacement units for stock, specify aluminum electrolytics if you want to keep your replacement parts stock in the "miniature" category.

PORTABLE STEREO 8 TAPE PLAYER TROUBLESHOOTING CHART

Symptom	Defect
No pilot light, no tape motion, and no amplifier hiss	Defective power supply cords Poor power connections Defective power transformer Defective ac-dc selector switch Defective auto—boat jack Shorted or open rectifier Defective cartridge switch
No operation with dc power source	Defective ac-dc selector Open auto—boat jack Broken or open dc power cord Defective batteries Improper seating of batteries Reverse-voltage diode Diode shorted or open
Pilot light on—no tape motion	Broken, loose, or worn drive belt Loose motor pulley Frozen or burned out motor Frozen capstan—flywheel assembly Open wiring to motor circuits
Wow or flutter	Poor lubrication Defective or oily capstan Loose or worn drive belt Oil on drive belt Belt riding high on motor pulley Defective motor Defective tape cartridge
Crosstalk	Height adjustment off Defective tape cartridge Jammed tape head assembly
Excessive noise	Worn or dry motor bearing Seized or frozen capstan Defective motor Erratic or worn volume control
Will not change channels	Shorted or open solenoid circuits Shorted diode across solenoid Open selector switch Defective track-shift mechanism

Symptom	Defect
Channel indicator defects	Open channel lamp circuit
	Defective channel lamp switch
	Burned out lamps
	Broken indicator lamps
	Slipping shaft to lamp switch
	Dirty lamp switch contacts
	Broken or defective rubber channel belt
No sound	Defective transistors
	Defective tape head
	Leaky or open coupling capacitor
	Shorted bypass capacitors
	Open B+ circuit to amplifier
	Defective tape
	Defective speaker or wiring
	Defective power cord
Weak or low sound	Defective or dirty tape head
	Defective transistors
	Open bypass or coupling capacitors
	Shorted bypass or coupling capacitors
	Tape head out of adjustment
Distorted sound	Leaky or shorted transistor
	Leaky coupling capacitors
	Burned or changed resistance of bias and emitter resistors
	Heatsink shorted to power transistors
	Very dirty tape head
	Open or erratic bias controls
Amplifier has high-pitched tone	Improper head azimuth adjustment
	Defective tape head
	Open capacitors in preamp
	Open or poor lead connections
Low treble	Oxide deposits on head
	Worn head surface
	Head out of alignment
Abnormal oscillations	Poor head leads
	Filter and decoupling filters
	Resistance change or burned decoupling resistors
	Open or poor volume control connections

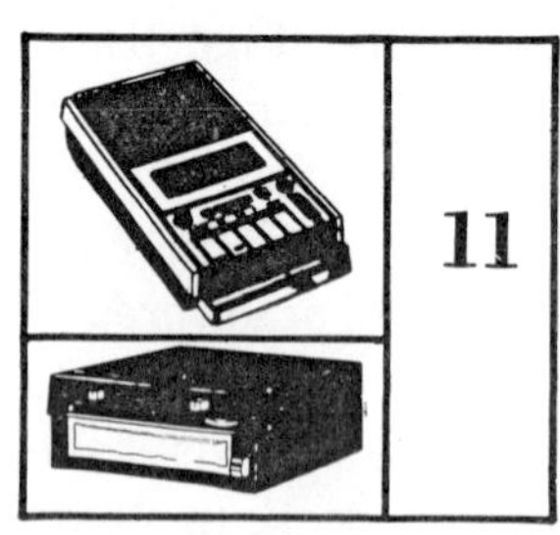

11 Discrete Decks, Combo Players, and 4-Channel Systems

The tape *deck* differs from the *player* in that the deck is played through the facilities of an existing stereo system. You simply plug the output of the deck into the tape or phono jacks at the rear of the cabinet. Some stereo consoles have a *tape* position included in the function selector switch. Today cartridge and cassette tape decks are found in a variety of entertainment consoles. The unit pictured in Fig. 11-1 is built into the chassis of an FM–TV console.

DISCRETE TAPE DECKS

A tape deck circuit is shown in Fig. 11-2. Note the absence of power amplifiers, speakers, and other "output" circuitry.

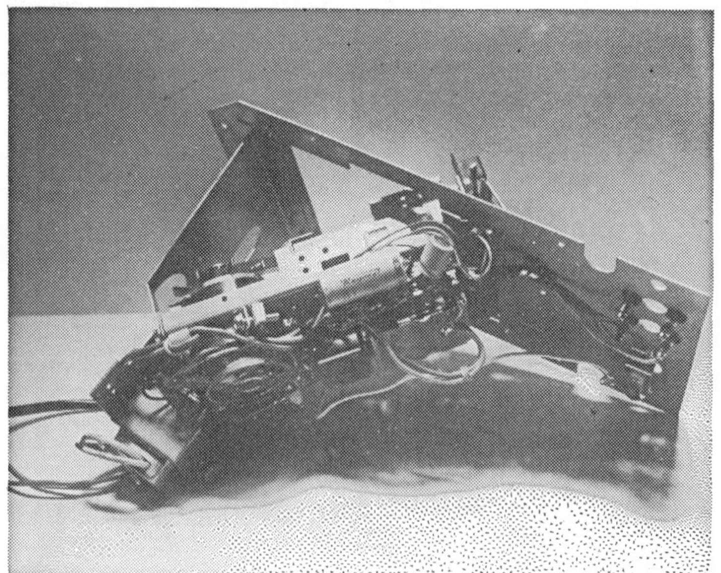

Fig. 11-1. The tape deck may be found in stereo consoles and TV color receivers.

The tape head is coupled to a transistor preamp stage. Some of the later tape deck preamps (such as the RCA unit of Fig. 11-3) include ICs rather than discrete transistors. In the better units, preamp volume may be set internally (RV701 in Fig. 11-3). The tape volume can either be raised or lowered to set the tape deck's level to match the levels of other console components (FM radio, phono, etc.). Some models have separate volume controls in the preamp tape deck; in others the only level control is that obtainable by adjustment of the main amplifier controls of the stereo player.

The preamp tape deck typically has its own ac motor, capstan drive assembly, track-shift solenoid, and track indicator lights. Generally, the preamp tape deck has all components mounted on one skeleton-like chassis and can be inserted into many different stereo units (Fig. 11-4). For service, the tape deck may be unplugged and removed from the cabinet. You can test the tape deck by plugging it into any stereo sound system.

Tape Head Alignment With Scope

You can make head adjustments on any tape player with a scope. Two adjustments are required to properly orient the

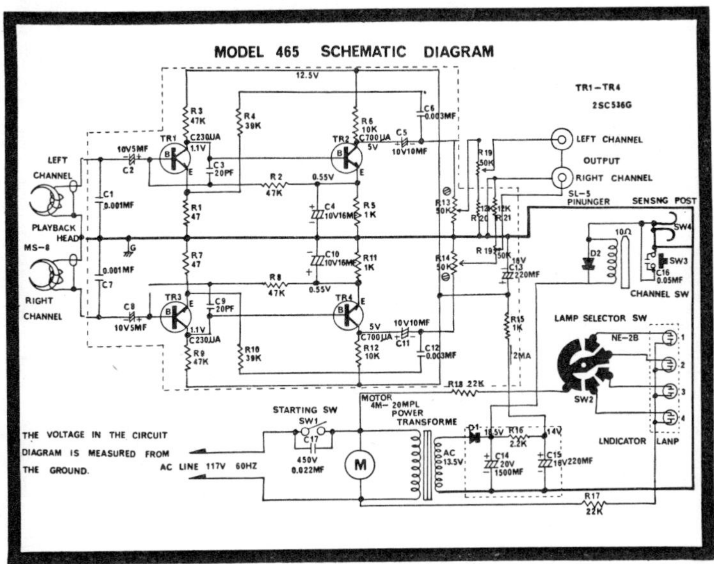

Fig. 11-2. Tape deck preamp circuit.

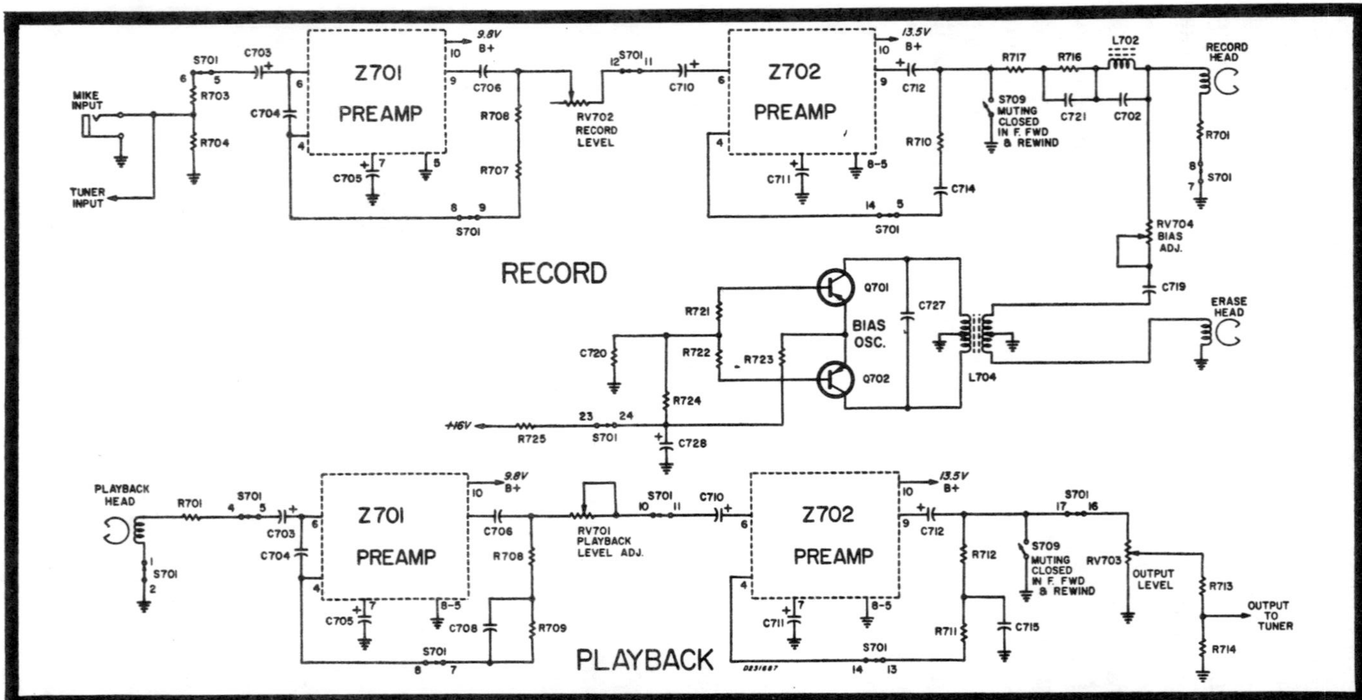

Fig. 11-3. This modern RCA tape deck includes ICs for preamplification and bipolar transistors for the bias oscillator.

tape playback head. Normally, the adjustments are correctly set at the factory and should not be changed unless the playback head or its mounting parts have been replaced. If someone has tampered with the head adjustments or jammed the tape head assembly, of course, alignment will be necessary. Head adjustment is required in units where crosstalk, noise, poor frequency response are the principal symptoms.

To make head height adjustment, load the instrument with a test cartridge (RCA 321). Set the track selector to program 2. Connect the scope to the left-channel output jack. Rotate the head height adjustment screw to find a null. (You can monitor the audio on system speakers by paralleling the scope with the amplifier input jack on the console.)

You can make azimuth adjustments with the same test cartridge. Monitor the right channel output with the oscilloscope. If the azimuth screw has a locknut, loosen it. Adjust the azimuth screw for maximum output from the right

Fig. 11-4. Most tape decks are mountable on a chassis assembly; they are skeletal in construction and can be inserted into many different models of console.

channel. You may have to repeat the head height and azimuth adjustments until a minimum indication for height occurs coincidentally with a maximum indication for azimuth.

You can double-check for crosstalk by inserting a different test tape (RCA 328) in the player. Push the manual shift program switch and check for crosstalk in programs 2 and 4. If crosstalk is noticed repeat head height adjustment or touch up the adjustment of the height screw. If erratic tape head movement is found when track-shifting, crosstalk may be noticed after several complete track rotations. Double-check by manually shifting the head several times. If you have a special short-length test tape (that switches programs every few minutes), plug it in and allow the player to cycle for an hour or so, then retest.

AC Motor Problems

Most transport motors found in ac-operated tape decks are *syncronous* types. They generally have three color-coded leads extending from the motor, as shown in the lower right corner of Fig. 11-5. The colored leads are yellow, black or blue, and red. A 1 or 1.5 μF motor starting capacitor is connected in series with the yellow lead.

The most common trouble with ac motors is dead circuit or intermittent rotation. If the motor will not rotate, remove the belt and turn the motor pulley with your fingers. See if the track indicator lights and amplifier are operating. If the lights are on and the console amplifier is okay, the motor or its associated wiring is at fault. Measure the line voltage across the red and black motor terminals. If voltage is present, remove the ac plug and take a resistance reading of the motor winding. A low-resistance reading should be obtained between all three motor terminal connections. If not, one of the windings may be open.

If the motor windings are normal, check for an open resistor (330 ohms, 7 watts in Fig. 11-5). Intermittent rotation can be caused by a defective starting capacitor, a shorted or poor winding connection inside the motor, or a break in one of the power leads. The motor should be replaced when overheating or intermittent rotation is noted.

Tape Stops Playing

The tape stops playing—this is a typical complaint found in the service report. For example, let's take the case of an Electrophonic Model T-500A. This stereo tape player also includes an FM multiplex radio chassis.

To rule out the stereo amplifier as the trouble source, switch to FM multiplex radio position. Turn up the sound and you should hear a loud rush indicating amplifiers are okay (cartridge inserted). If you hear the rush but the motor of the deck still does not operate, remove the playback deck from cabinet for examination. Check to see if belt is off the flywheel. Check also for a defective cartridge switch. Check the motor and connections.

If the belt is off and the motor is rotating, check for a loose, oily, damaged, or stretched belt. (In some models an adjustment can be made to tighten the belt.)

Clean up the motor pulley, flywheel, and capstan drive.

Clean up the tape head and automatic switch tongs.

Check the automatic and manual channel switch operation.

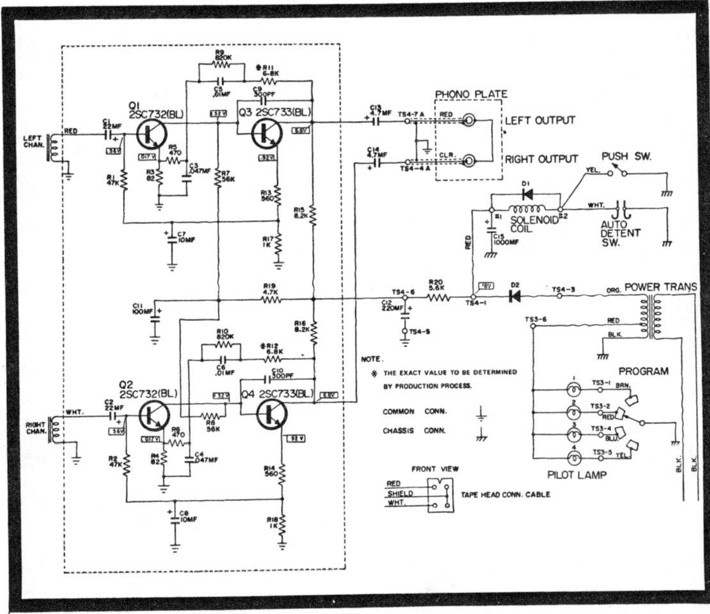

Fig. 11-5. Model TDA302 tape deck circuitry. Note motor connections and color-coded wiring.

As part of your routine service procedure, check for crosstalk and make necessary height and azimuth adjustments. Clean and demagnetize the tape head. Let the unit play a few hours and notice any speed changes or erratic channel switch operation.

Make sure the radio and phonograph sections are functioning before releasing the tape player. If everything checks out, clean up the cabinet and metal front assembly. Clean off all knobs and switches.

AMPLIFIERS IN COMBO SYSTEMS

The problems in combo-system amplifiers are no different than those typical of other amplifiers. However, there is one type of amplifier we haven't discussed yet—the direct-coupled output stage, which uses no transformer or capacitor to couple the amplified signal to the loudspeakers. Figure 11-6 shows one such circuit. It should be noted that this amplifier configuration offers excellent performance when it's working properly. But when trouble develops in the dc output circuit, there is nothing to prevent a dc voltage from being applied directly to the voice coil windings, damaging the speaker.

If the complaint on a combo amplifier is a dead speaker or earphone, suspect problems in the dc output circuits. It's possible to have a defective speaker or earphone, of course; but if both are not working (on one channel), suspect the amplifier circuits. It's unlikely that both components are defective. You may quickly damage a test speaker or earphone if it's plugged into a defective dc output circuit, so exercise care: keep the volume low and monitor the output with a dc voltmeter.

How do you know the amplifier has a dc output circuit? Only a few manufacturers to date are using this type of circuit. After servicing a few, you'll quickly recognize these models by the absence of large-value electrolytics or ungainly transformers near the power transistors. The presence of dc on the speaker voice coil is evidenced by a classic symptom: when the speaker is touched to its proper amplifier terminals, the cone will move forcefully either in or

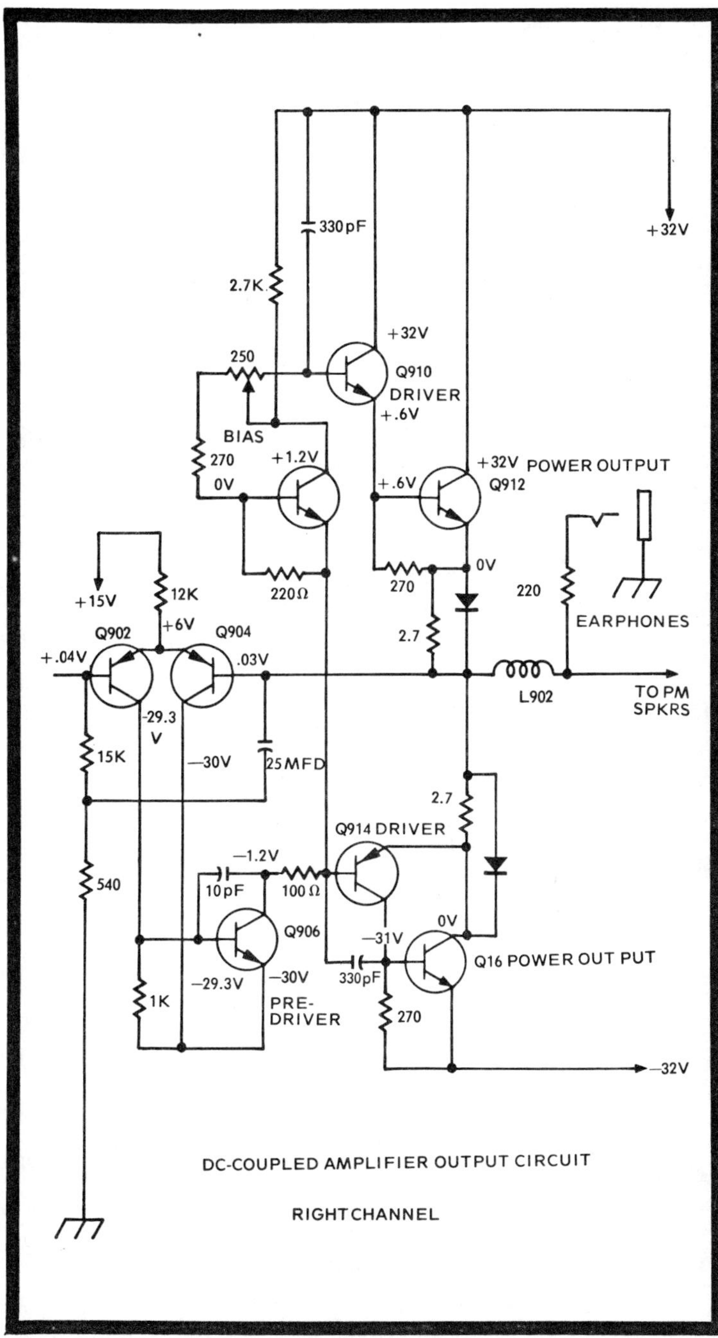

Fig. 11-6. DC-coupled amplifier output circuit.

out, the distance of travel determined by the magnitude of the voltage and the direction determined by voltage polarity.

A dc voltage found at the speaker output terminals is not necessarily an indication that the amplifier is a direct-coupled output circuit, but it is a sure indicator that something is amiss. It is possible to have dc voltage upon the speaker terminals with a leaky or shorted electrolytic speaker coupling capacitor, which is found in quasi-comp and comp-sym output stages.

To preserve your test speaker, the recommended service procedure is to use a resistor as a load for amplifier testing. Make up a couple of resistor loads (using 8—10 ohm resistors with 5 or 10W capability) for these checks and use them for all problems related to dc in the output stage. Use the same resistors for speaker loading while servicing the amplifier or in making tape head adjustments. Do not connect any speaker to the resistor. Just connect a voltmeter across the resistor and leave it connected while servicing the defective channel. Now you may service the dc output amplifier circuits without possible damage to the test speakers. When the dc voltage is removed from the load resistor, you have located the defect.

Most problems found in the dc output circuits are leaky or open transistors. Occasionally, a shorted or leaky diode will show up. Most burned resistors are caused by a shorted or leaky transistor. Check the power output, driver, and bias regulator transistor for leaky or open conditions. Failure of any one of these transistors in a direct-coupled circuit will produce dc voltage at the speaker terminals.

First, check the bias regulator and power output transistor, since these have proved to be the most troublesome. In some models, these transistors may be plugged into sockets or bolted to a chassis heatsink. Be careful when removing and installing these transistors; observe all plug-in and mounting procedures.

Troubleshooting IC Power Circuits

In many of the latest combo amplifier circuits, ICs are used as both preamp and power output. There are many types of power ICs found in various units. The photo of Fig.

11-7 shows but a few. The entire audio section may be incorporated into but one IC. The af preamp ICs are quite small physically. Large ICs usually imply a considerable amount of power-handling capability.

Sometimes a shorted or leaky IC power output stage will blow the amplifier fuse. When a slightly larger-than-normal fuse is clipped into place, you can touch a suspect IC in operation to assess its condition by temperature. At low volume levels it should not get warm. If it does, shut the amplifier off. In many cases visual inspection will quickly locate a defective power output IC. If the fuse blows, measure the current drain in that circuit. A current meter inserted into the dc power voltage line to the suspect IC will verify an IC problem.

With signal injection, start at the input terminal with an audio signal generator. Connect the scope leads across the load resistor. The trace should show your applied signal riding on a dc level. Remember, the shorted or leaky power IC runs quite warm. Touch it and compare with the like component in the opposite channel.

Removing the power IC may take a little longer than removal of a conventional small-signal IC, since it will

Fig. 11-7. Integrated circuits of modern vintage can carry considerable current. Most have a telltale heatsink as part of the integral package.

Fig. 11-8. In this chassis the power IC is bolted to a metal flange.

probably be bolted to a chassis-mounted heatsink. In Fig. 11-8 the power ICs are bolted to a metal flange that connects to the chassis. First, remove the bolts and nuts securing the IC. A solder-sucking iron or Solder-Wik will quickly remove excess solder from the PC boards. Be sure all solder is removed before attempting to lift the IC from its mounting.

With the power IC shown in the photo, there are 10 soldered connections. You may cut off each terminal with side cutters and unsolder each lead; but this takes time. Also, it renders an IC valueless, which is particularly bad if you've incorrectly traced a problem to the IC. Careful desoldering is the best bet. The circuit may be lifted from its position with a knife or screwdriver blade under the component. There are several IC clips available that clamp over the ends of the IC and quickly remove it. (You do not have to be as careful in removing the defective component as installing a new one, of course. But tearing up a suspect IC is like burning a bridge. If you've guessed wrong about the route, you can't go back.)

When you install the new power IC, be sure all pins are straight and in line. If a removal clip is not handy, start by placing one side of the chip into position first. Once that side is partially secured, move to the opposite side. Be careful not to bend any one of the pin connections under the unit. Double-check all connections on the PC wiring side. Scrape and clean off all excess rosin with a knife or stiff wire brush.

Audio Generator Testing

When difficult distortion problems are noted an audio and generator with sine- and square-wave capability will prove helpful in locating the defective stage. By going from stage to stage, even the slightest amount of distortion may be located by observance of the waveform on a scope. Always keep the signal generator output low enough not to overdrive the transistor stages. Check the audio signal generator settings on the known-good stereo channel and inject this same signal into the distorted channel. With an applied sine wave, distortion most often occurs at the top or bottom of the sine wave. If distortion occurs at both the top and the bottom of the waveform, an amplifier stage is being overdriven. Correct this by lowering the waveform's signal level. (See Fig. 11-9.)

The square wave is a particularly useful waveform in checking amplifier problems, because the character of the reproduced square wave will give immediate clues as to the nature of the problem.

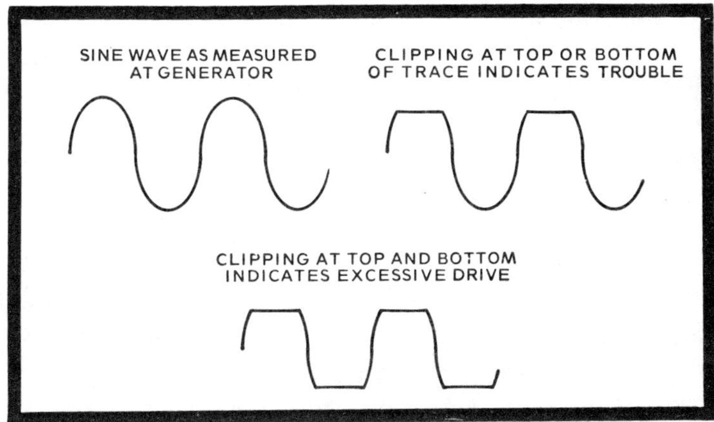

Fig. 11-9. The sine wave should look the same at the speaker terminals as it does at the generator.

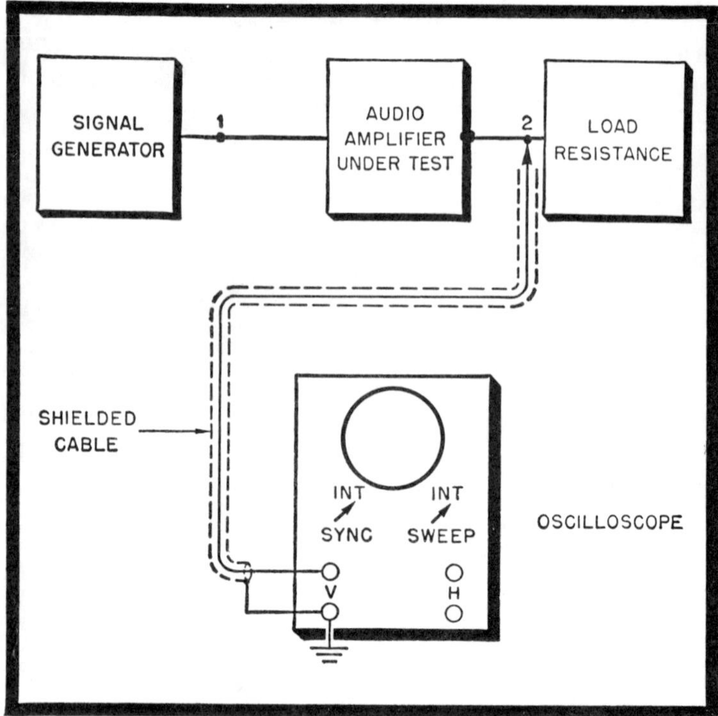

Fig. 11-10. Basic test setup for audio amplifier testing.

The basic test setup for square-wave testing is shown in Fig. 11-10. A square-wave generator is substituted for the sine-wave generator. For the testing of any circuit, the signal generator with its normal load, or a device equivalent to its normal load, should be inserted where the audio amplifier and load are shown in the illustration. If an audio amplifier is tested without its usual load, a resistor should be inserted for the substitute load. The frequency range of a typical generator used for these tests is from 20 to 10,000 Hz, with an output impedance of 500 ohms.

A square-wave signal of constant amplitude should be fed to the input of the circuit under test. The oscilloscope should then be placed across the input of the circuit to make certain that the oscilloscope itself responds to the waveform. If the oscilloscope does not reproduce the generated waveform perfectly, the actual waveform it does give can be used for comparison with the output. If the circuit tested consists of more than one stage, the oscilloscope probe should be moved from stage to stage in succession, to check the output of each stage against its input.

In square-wave testing with the oscilloscope, it is particularly important that the vertical deflection amplifier in the scope have a frequency-compensated attenuator. Figure 11-11 shows the effect of feeding a 15 kHz square-wave signal to the vertical amplifier of an oscilloscope through an uncompensated attenuator. Here it is shown that, although the vertical amplifier of the oscilloscope is rate to pass a 15 kHz square wave without distortion, the attenuator cuts down on the frequency range for which measurements can be made with the oscilloscope. In such cases, the input signal to the oscilloscope should be fed directly to the vertical deflection plate terminals of the oscilloscope.

Response Checking of Circuits. The response characteristics of most circuits are limited by the lowest and the highest frequencies that the circuit passes. The high-frequency transmission determines the shape of the transient at the instants that the signal is applied and removed. The low-frequency transmission of the circuit determines the value and the shape of the waveform after some time has elapsed (between the application and the removal of the square wave for each half-cycle). If a square wave is fed to a circuit and the output wave is rounded off as in B, Fig. 11-11, it indicates that the response of the circuit to high frequencies is poor. If the output waveform has sharp

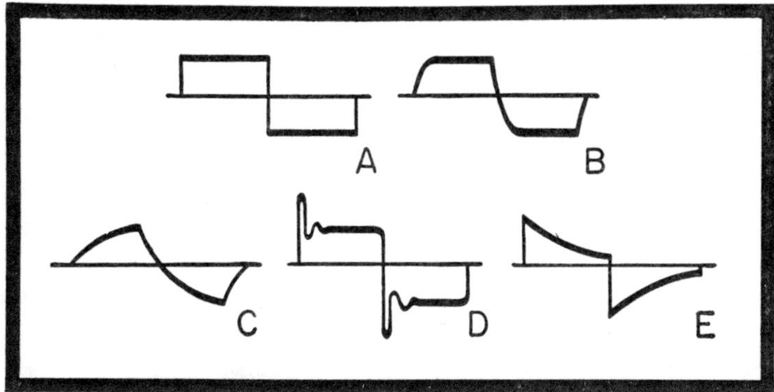

Fig. 11-11. Square-wave input to amplifier and typical output patterns.

corners, however, and the center of the waveform plateau has a dip or slant as in E, the low-frequency response of the circuit is poor. This indicates that the circuit passes the high-frequency components of the input waveform and only a limited portion of the low-frequency components. C shows the response of a circuit having poor high-frequency response.

Checking Audio Amplifiers. Because the square wave is composed of a fundamental frequency and a large number of odd harmonics, passing a square wave through a circuit immediately gives information about the frequency response of the circuit to a wide band of frequencies. The square waves generated by most generators contain the fundamental and about ten odd harmonics of appreciable amplitude. (Actually, the square wave contains many more harmonics, but their contribution to the shape of the wave is negligible.) If an amplifier cannot pass any of the harmonic frequency components of the square wave, the output waveform will not be square, and its shape will depend on the remaining harmonics. To cover the frequency response of a circuit, therefore, it is necessary to use only two different frequency settings on the square-wave generator. To test the frequency response of an audio amplifier from 50 to 10,000 Hz, a 50 Hz square wave should first be fed to the circuit. This tests the low-frequency response of the amplifier. If the amplifier passes equally all frequencies from 50 to about 400 Hz, the output waveform will be square; that is, there will be no appreciable difference between it and the input waveform. To test the high-frequency response of the circuit, the signal generator should be set to about 1 kHz. If the circuit amplifies all frequencies equally up through the tenth-odd harmonic (that is, up to 20 kHz) the square-wave output should be exactly like the input. This would indicate good response in the upper audio-frequency range from 1 to 20 kHz. Besides furnishing information on frequency response, square-wave testing of an audio amplifier uncovers a great variety of circuit defects. If the output waveform at any stage exhibits a thickening of the flat-top portion of the wave, it indicates the presence of hum voltage superimposed on the signal. Oscillation in the high-frequency range is shown by an output waveform like that shown in Fig. 11-11D. If oscillation occurs in the low-frequency range of the amplifier, the pips occur nearer the right end of the waveform plateau. Figure 11-12 shows the most common types of distortion with low- and high-frequency square-wave inputs.

Output Transistor Bias Adjustment

The output transistor bias adjustment should be made when distortion is occurring or when the output transistors are replaced. Sometimes misadjustment will cause low-level output or crossover distortion. These bias adjustments are typically screwdriver controls and are located in the emitter circuits of the output transistors. (These bias controls are only found in high-wattage output amplifiers.)

First, let the set warm up and leave the volume control at minimum position. In some amplifiers, the manufacturer will give the exact voltage measurements to be observed at the emitter of the output transistor. This voltage is generally equal to one-half of the collector voltage. The bias adjustment is rotated until the required emitter voltage is

INPUT TO AMPLIFIER	SHAPE OF OUTPUT WAVE SEEN ON CRT SCREEN	INTERPRETATION
50 Hz SQUARE WAVE		GOOD LOW-FREQUENCY RESPONSE AND NEGLIGIBLE PHASE SHIFT
		LEADING LOW-FREQUENCY PHASE SHIFT AND LOW-FREQUENCY ATTENUATION
		LAGGING LOW-FREQUENCY PHASE SHIFT
1000 Hz SQUARE WAVE		GOOD HIGH-FREQUENCY AND TRANSIENT RESPONSE
		POOR HIGH-FREQUENCY RESPONSE
		EXCESSIVE HIGH-FREQUENCY RESPONSE (OSCILLATION) AND NONLINEAR TIME DELAY
		EXCESSIVE OR INSUFFICIENT MIDFREQUENCY RESPONSE AND NONLINEAR TIME DELAY

Fig. 11-12. Output waveforms from audio amplifiers with a square-wave input, showing defects causing distortion.

obtained. You should make this bias adjustment in both channels.

The best method is to feed the audio signal at the base of the driver transistor. Connect a scope at the audio output terminals. Increase the audio signal from the generator until clipping is noticed. Now, adjust the bias control for equal clipping on the top and bottom of audio signal with minimum crossover distortion and equal channel output.

Sometimes the bias adjustment control may produce a noisy condition. The center wiping contact becomes dirty or corroded and makes a poor connection. Sometimes noise will exist at this connection and produce popping sounds in the amplifier section. Replacement of the bias control will solve the noisy condition.

4-CHANNEL PROBLEMS

Four-channel problems are no different from those found in a conventional stereo amplifier. There are four separate channels, but service problems on an individual-channel basis are the same. There are more components to break down and they are mounted a lot closer together. Locating the amplifier components is the most difficult problem in any 4-channel circuit.

The 4-channel tape head may cause more problems than the standard stereo head. Remember, there are four in-line windings within a single tape head. Movement of the tape head may produce cable problems at the head terminals. When one channel appears dead or intermittent, check first for a loose or broken lead (or plug) at the tape head. You may inject the audio signal at each tape head connection to determine if each amplifier is functioning. Four-channel tape head alignment is essentially the same as in any stereo player—but it's a bit more difficult to eliminate crosstalk.

When the 4-channel player will not shift to all four program positions, suspect a defective actuator switch. Check to see if all contacts are made on the channel switch assembly. A set of contacts may be bent out of line or become dirty. See if the solenoid movement is normal at the relay switch. Check for a shorted suppressor diode across the relay winding. Also, check for broken wiring connections at switch and relay assembly.

A component layout diagram is almost a must in locating defective parts. In some 4-channel circuits the audio output transistors may be positioned side by side. Generally, the two left channels are mounted on one board assembly, and the two right-side output circuits are mounted on another. Without a component layout chart, it may be very difficult to locate the defective output transistors.

When a dead or intermittent channel is noted, you can locate the defective stage with another audio amplifier. Insert the 4-channel tape cartridge and start at the speaker output terminals. If a component layout chart is not available, check the output from each set of power output transistors. Remember, these power transistors tend to come in pairs. Touch the metal case of each power transistor. One power transistor may be insulated above ground and the corresponding transistor mounted directly to chassis, although in some instances you will find both transistors insulated above chassis ground. Compare the signal of each power transistor as you proceed in checking each audio transistor. When no signal is noted or appears intermittent, you have located the defective channel. You may also use the scope to quickly locate the right set of power output transistors.

AUXILIARY INPUTS TO COMBO AMPS

In the larger combo units, a separate jack is provided to play a guitar or other external audio source through stereo amplifier. Figure 11-13 shows a guitar preamplifier stage as incorporated in *Lloyds* Model 3M37 unit. This small preamp stage is found on the main chassis close to the preamp audio circuits. The circuit consists of a single preamp transistor stage.

When the guitar section appears dead or intermittent, double-check to see if the tape and radio sections are functioning. Check the guitar cords and plugs for possible breakage or poor connections. If the problem is tied to the preamp stage you may signal-trace the circuit with the audio

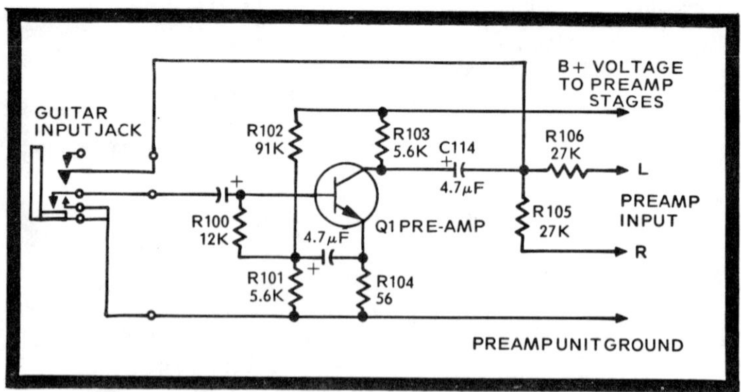

Fig. 11-13. Lloyds 3M37 model guitar preamp stage.

signal generator. You can quickly check the transistor by injecting signal at the collector and then the terminal base of the preamp transistor. With the volume wide open, touch the small preamp transistor. You should hear a loud pickup hum. If not, check the transistor within the circuit. A loud audio signal at the collector terminal indicates the amplifier is good. No or low audio signal at the base of the transistor indicates a defective preamp transistor.

CONSOLE & COMBO TROUBLESHOOTING CHART

Circuit	Symptom	Probable Cause
AM	No sound	Defective power switch Defective selector switch Defective demodulator diodes (Isolate problem to AM section by checking operation of FM section.)
	Weak reception	Poor antenna coil connection Rf transistor open or leaky Poor rf tracking I-f alignment off
FM	No sound	Defective selector switch Defective demodulator diodes Open circuit of antenna or FM oscillator coil

Circuit	Symptom	Probable Cause
FM mono	Weak reception	Bad rf transistor Bad mixer stage Rf tracking off Poor soldered contacts in FM section Bad rf tracking I-f tuning off
FM stereo	No sound	(Isolate to mpx section by checking FM mono reception.) Defective selector switch Bad FM mpx transistor
	Poor separation	Transistors out in multiplex circuit Bad FM diodes Shorted turns, 19 kHz, 38 kHz amp and 38 kHz coils Improper alignment
	No stereo	Indicator lamp or circuit problems
	Indicator light	Multiplex alignment problems
Preamplifier	No sound	Open or shorted tape head Defective transistors Defective selector switch Leaky or open coupling and bypass capacitors No voltage to preamp section
	Poor sound and weak volume	Deposits on tape head Preamp transistors bad Open coupling electrolytics
	High-pitched tone	Improper head azimuth angle Defective tape head Incorrect capacitance or open across tape head
	Low-pitched tone	Reduced or leaky capacitance in first preamp emitter circuit

Circuit	Symptom	Probable Cause
		Open or change of resistance in feedback circuitry
	Distorted output	Change of resistance of collector load and bias resistors Leaky coupling and bypass electrolytics
	Abnormal oscillations	Defective tape head Low capacitance of preamp collector bypass capacitor Resistance change of decoupling resistor Poor filtering
Solenoid circuit	Poor or no head switching	Defective channel switch Shorted suppressor diode across solenoid Open solenoid winding Shorted turns of solenoid

Case Histories

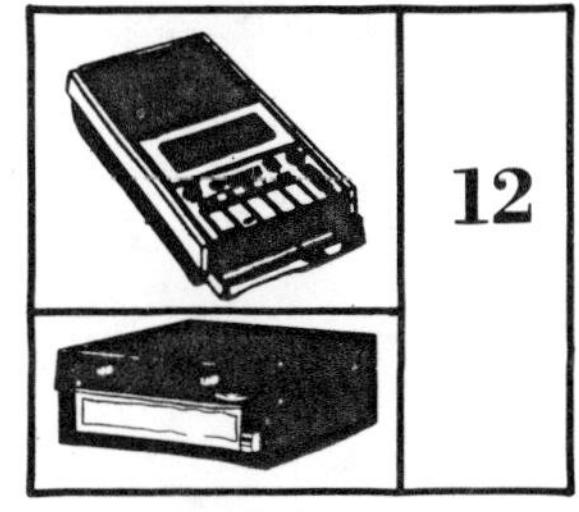

This chapter deals with actual case histories of problems related to cassette and cartridge players. These problems are broken down into these three categories: cassette players, auto-mounting cartridge players, and ac-operated "home" systems. Entries are by make, model, and problem. Although these problems deal with specific models, the method of locating and servicing applies to any cassette or cartridge player.

CASSETTE PLAYER PROBLEMS

Make & Model: *Admiral Model CTR 450*

Symptom: *Play button won't lock down.*

In this model the play button would not lock down and the motor would not rotate. This lock problem was caused by a soft metal lever that was easily bent out of line. The small button catch assembly was bent to clear the bottom lever and lock it into position.

When in play position the power switch contacts were open. These same contacts are made in record and fast-forward, but a different shaft is used in play position. The switch leaf was bent back in place to operate in play position.

Make & Model: *Ampex Micro 85*

Symptom: *Tearing tapes.*

Both tape head and pressure roller were packed with oxide tape residue. Several pieces of tape were removed from around the capstan drive and pressure roller. A flat plastic rod was used to remove packed oxide from tape head. The record–playback and erase heads, tape guides, and rubber rollers were cleaned with alcohol. A drop of light oil was applied to ends of pressure roller and capstan bearing.

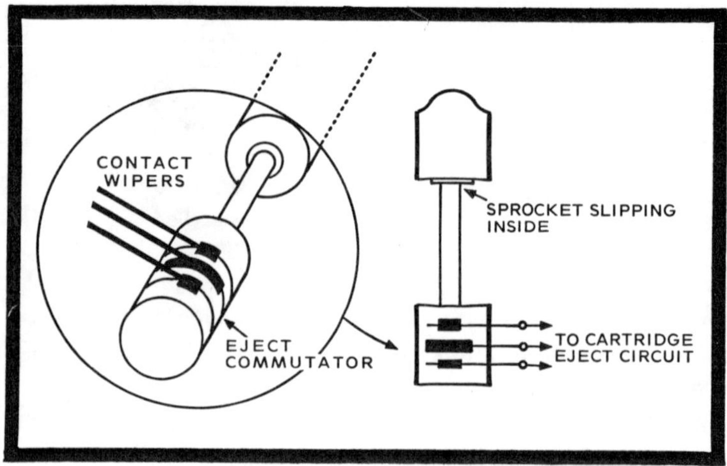

Fig. 12-1. If the eject commutator, connected to one of the reel hubs, does not turn, cassette will eject. The commutator typically has three printed conductor patterns. Each must make good contact with the leaf springs.

The tape head was demagnetized with a cassette pencil-type unit.

Make & Model: *Automatic Radio Model 9956*

Symptom: *Keeps ejecting cassettes.*

This auto stereo cassette tape player would start to play, but would eject within seconds. Also, it would not operate in rewind position. The trouble could be a defective pulse switch, electronic eject circuit, or solenoid. But since the player would not rewind, the trouble was isolated to the supply spindle that is common to both rewind and pulse switch commutator (Fig. 12-1).

The small sprocket that fits inside tape cassette was missing. This let the small supply spindle shaft slip inside and not turn commutator pulse switch, ejecting cassette. Also, in rewind position the supply spindle would not rotate. Replacing the small spindle sprocket cured both problems.

Make & Model: *Elgin Model R4500*

Symptom: *Runs too slow.*

After removing the bottom cover it was easy to see that the motor drive belt had slipped down over the belt guide. The small belt was removed and cleaned. Capstan and motor drive pulley were cleaned off with alcohol and soft cloth. A drop of oil was placed on capstan bearing. Also, complete cleanup of tape head and pressure roller solved the slow-speed problem.

Make & Model: *Elgin Model R4500*

Symptom: *Noisy channel.*

It was found when the amplifier chassis was moved or twisted the noise would come and go. All chassis screws were tightened and seemed to eliminate the noisy condition. Sometimes it is wise to tie or ground the amplifier to motor and common chassis ground to eliminate possible poor grounding or excessive motor noise that is picked up in amplifier.

Make & Model: *Holliday Model 16-038*

Symptom: *Amplifier noise—no tape motion.*

The characteristic rushing noise indicated that the power switch and amplifier were operating. The supply and takeup spindle would not rotate. Either the belt was off, the motor was defective, or power was not getting to the motor.

After removing outside covers the motor belt was removed and motor pulley rotated with fingers. The motor "wanted" to move but was stuck. The motor was removed and dismantled. Motor bearings and armature shaft were cleaned with alcohol. A drop of light oil was placed on each end bearing. (Replace motor if bearings are worn and noisy.)

Make & Model: *Lloyds Model 1V92A*

Symptom: *Defective takeup spindle.*

Within this cassette player the tape would bunch and not wind up properly. The takeup hub seemed to be slipping. Another cassette was inserted with the same results. Generally, a fiber or cork friction drive system is found between plastic pulley and spindle drive assembly (Fig. 12-2). In this case,the friction drive on takeup pulley was slipping. The friction drive assembly was removed and cleaned. Greater spring tension was made by physically manipulating the spring.

Make & Model: *Panasonic Model RQ226S*

Symptom: *Extreme distortion.*

This cassette player had distortion in play and record position. A new cassette was inserted, with same results. The distortion had to be in the amplifier section common to *play*

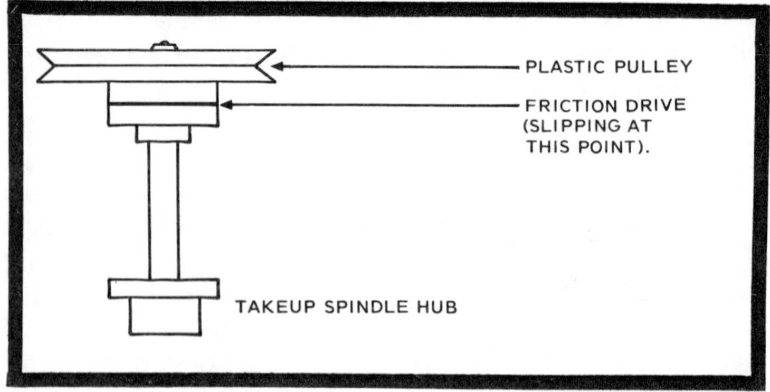

Fig. 12-2. Lloyds model 1V92A-114 casstee would bunch up tape; slipping occurred at friction drive assembly.

and *record* operations. Since distortion was projuced in the audio output stages, they were checked first. Upon removing the metal heatsink screw it was obvious that the metal piece had been biting into the insulation of terminal leads. The transistor leads were shorted to ground through the metal heatsink. Redressing insulation and terminal leads of the output transistors solved the distortion problem.

Make & Model: *PennCrest Model 3240*

Symptom: *Will not record; playback distorted.*

A new cassette was tried in this cassette player to verify the complaint. The symptoms were still there. Weak volume was noted in playback.

The tape heads were checked and found caked with oxide dust. A good cleanup with alcohol cured the condition. It's wise to check the tape head when weak audio or distortion problems are found before removing cassette assembly from the plastic cabinet.

Make & Model: *RCA YLS 15B*

Symptom: *Rewind okay—no play motion.*

This portable cassette player would neither rewind nor operate in record or play modes. The supply spindle was ruled out as it operated in all three modes. Either the takeup spindle or capstan–flywheel assembly was at fault. Sometimes the flywheel would turn and then stop.

At first, after rotating the flywheel by hand it was found to be either gummed up or frozen in play operation. When not in play or record mode the flywheel drive seemed normal. A closer inspection revealed a dry and frozen rubber pressure roller was the problem. A cleanup and lubrication of the pressure roller bearing placed the cassette player in operation. Always clean and demagnetize the tape heads at the same time.

Make & Model: *RCA Model YJD 165*

Symptom: *Ruining tapes.*

The complaint of this cassette player was destroying and pulling cassette tapes. Cleaning up the capstan drive, tape head, and pressure roller did not solve the problem. The cassette player was operated upside down without cassette. Someone had already tampered with and attempted to repair the tape recorder. After checking with physical layout and manufacturer's schematic, I could see the drive belt had been installed on the wrong side of the flywheel. Installing the rubber drive belt on the correct side of flywheel solved the rolling up and pulling tape from the cassette cartridge.

Make & Model: *RCA Model YJD 165*

Symptom: *Won't function without remote plugged in.*

This cassette player would not play with the remote control out of its socket, and the remote switch would not shut off the tape player. At first the power switch was suspected of being defective. After checking the remote female plug with remote cable removed, the long pin of the female plug was seen to be shorting out the remote switch. The plug tip would push up against the center pin of the female plug, turning the player on. Alignment of female plug cured the problem. In many cases the remote and earphone jacks can be thrown out of line, leaving open contacts. A quick check of these jack contacts can solve many cassette operation problems.

Make & Model: *RCA Model YZD 165*

Symptom: *Play but no record.*

The recording meter would not indicate when in record mode. Another microphone was plugged in with the same results. The amplifier, playback head, and microphone checked out good. This left a possible defective record switch or wiring. The record–play switch was sprayed for dirty contacts. Only a weak recording was noted. Soldering all 18

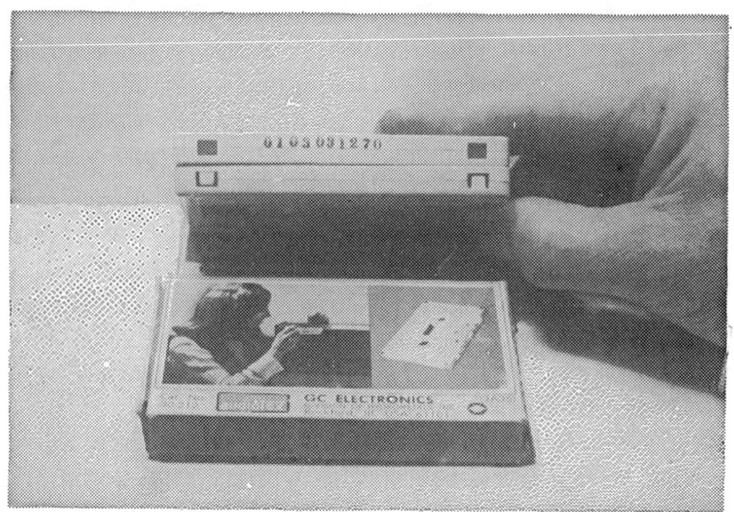

Fig. 12-3. When tabs are removed in the back of the cassette (top) the player will not record. Notice the tabs are still in the bottom cassette, and this cassette will record.

contacts on the record–play switch solved the no-record condition.

Make & Model: *RCA Model YZB 508*

Symptom: *Noisy.*

This battery-operated cassette player had a noise in the amplifier in record or play position. The motor belt was removed and the noise disappeared. When the motor was operating with belt in place the noise appeared once again. The motor belt was removed and motor placed under load by slowing the motor pulley between two fingers. Undoubtedly, the noise was produced from arcing motor brushes. The motor was replaced.

Make & Model: *RCA Model YZB 508*

Symptom: *Crosstalk—no erase.*

The output of this cassette player was a garbled mess. A new recorded cassette was played and seemed normal. In record position the circuit would not erase the previous recording. Removing the chassis revealed a dc magnet as erase head. In record position, the erase magnet would not press against the moving tape. Bending the dc erase magnet holder cured the garbled recording.

Make & Model: *RCA Model YZB 520*

Symptom: *No record.*

The owner complained of *no record* with this small cassette recorder. A new tape recording was left in the player. The recorder played normal but would not record. Another cassette was placed in the recorder and recorded perfectly. Upon checking the suspected cassette I could see the small tabs on the rear of the cassette were removed (Fig. 12-3). Placing vinyl tape over the tab openings solved the problem.

Make & Model: *RCA Model YZB 520J*

Symtpom: *Will not play on batteries.*

The problem with this cassette player turned out quite simple. Ac operation was normal, but it simply would not operate on batteries. The operator left the ac cord plugged in and this model (like many others) will not operate on batteries *unless cord is removed.*

Make & Model: *RCA YZB 520J*

Symptom: *No record or playback.*

The rewind and fast-forward were normal in this cassette player, but there was no record or playback. When switched to playback mode the cassette seemed to bind. Closer inspection revealed the tape head was twisted and would not fit inside cassette. Someone had attempted to adjust the azimuth screw and tilted the tape head. Proper azimuth alignment cured the problem.

Make & Model: *RCA Model YZB 525 E*

Symptom: *Motorboating.*

When volume was turned up high the amplifier would start to oscillate, as evidenced by a motorboating sound. Most problems found in cassette players with oscillations can be caused by poor switch contacts or filter capacitors. Sometimes the switch contacts are dirty or worn so as not to make proper contact. This problem was a leaky 10 μF, 16V decoupling electrolytic capacitor. Small electrolytics should be clipped across the suspected capacitor with player turned off.

Make & Model: *Ross Model 8600*

Symptom: *Remote would not shut off.*

Operation was normal when cassette player was played without its remote control mike. With remote plugged into unit, the player would not shut off. The remote switch was suspected of being defective. When checked with ohmmeter the remote switch was found okay. The problem turned out to be a broken wafer insulator of remote female plug. Some of these remote, earphone, and auxiliary jacks can be replaced separately. In this model, all three jacks were mounted on one assembly; so the assembly could be replaced with the exact part number.

Make & Model: *Sony Model TC910*

Symptom: *Dead set.*

Not a sound or movement was heard from this cassette player. The trouble had to be in the power supply. Both silicon diodes were shorted and replaced. The small player was still dead. The ac voltage was measured on primary and secondary windings of small power transformer. Ac line voltage was found on the primary and no voltage on the secondary windings. The ohmmeter found an open primary winding (Fig. 12-6). Since this transformer had several winding taps, the exact replacement part was ordered and installed.

Make & Model: *Soundesign Model 7843*

Symptom: *Sluggish fast forward.*

Sometimes the fast-forward speed was intermittent and erratic. Play and rewind speeds were normal. In this model there are two separate motors—one for record and playback mode. The other motor operates rewind and fast forward. Since rewind speed was good I assumed the R/FF motor was okay. The fast-forward motor belt was found greasy and bunched up against the chassis frame in operation. Removal of rubber belt and good cleanup with alcohol solved the sluggish fast-forward speed. If a motor drive belt appears worn or loose, replace it.

AUTO-MOUNTING CARTRIDGE PLAYERS

Make & Model: *Airline Model 16713B*

Symptom: *Only a rush noise, no other audio.*

I could hear a rush noise indicating the amplifier was working, but there was no music when a recorded tape was inserted. A pencil eraser end was used to push against the cartridge switch; there was no capstan movement. Removing the cover revealed a thrown motor belt. The capstan—flywheel would not turn freely and seemed dry.

Removing the capstan—flywheel revealed sticky and dry bearings. The capstan drive shaft and bearings were washed out with alcohol. Lubrication and a good cleanup of capstan assembly did the trick.

Make & Model: *Amco Model 4V9*

Symptom: *Won't automatically change channels.*

This auto stereo tape player would change channels with manual button, but not automatically. Either the automatic switch or connecting wire was defective. One side of the automatic prong-type tape switch was broken off (Fig. 12-4). Most of these automatic tape switches are built alike and can be substituted for one another. Keep a few on hand, as this problem is found quite often. A temporary repair can be made by drilling or grinding down the small rivets and installing a formed copper strip that resembles the original.

Make & Model: *AMCO Model 4V9*

Symptom: *Blows fuses when track-shift button is pushed.*

Another tape player of the same model would blow fuses when it changed channels. Generally, this condition is caused by a shorted silicon diode across track-shift solenoid (Fig. 12-5). Besides a shorted diode, part of the printed board was burned loose in this model. The burned wiring was removed

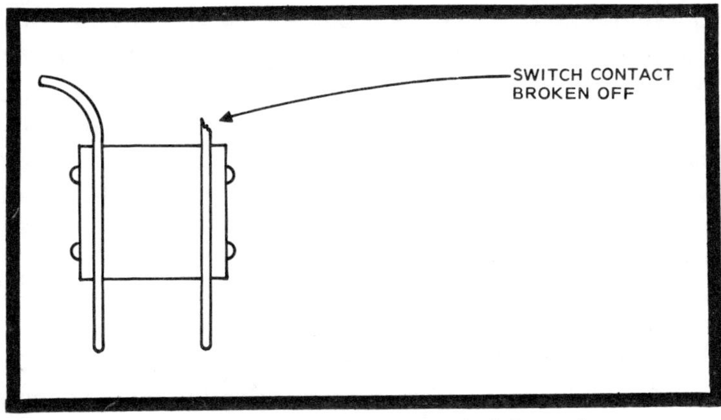

Fig. 12-4. Broken automatic track-shift switch.

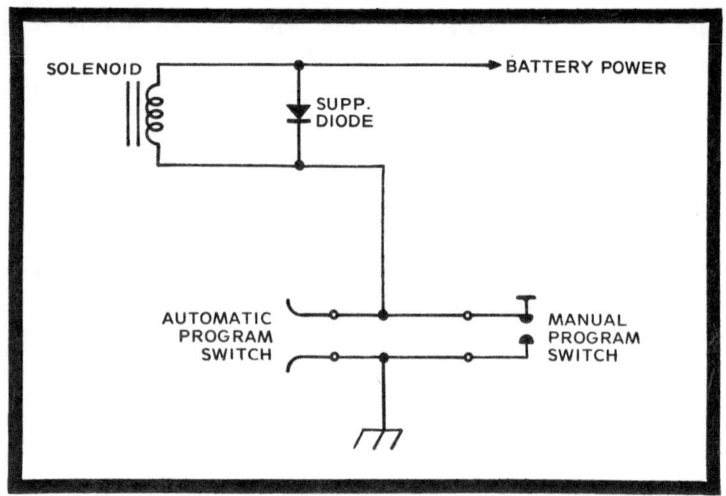

Fig. 12-5. A shorted diode blows fuses when solenoid circuit is energized.

from PC board and replaced with hookup wire. Be sure and observe correct polarity when replacing silicon diode.

Make & Model: *Automatic Radio Model CF6-6745*

Symptom: *Tears tapes; no sound.*

The unit played perfectly upon the service bench, although the complaint was *ruining tapes* and *no sound*. This tape player was pulled and brought into the shop by the customer. The player ran for two days without skipping a beat.

After no problems were indicated with the tape player, the complete tape head assembly was cleaned with alcohol and soft cloth. Actually, the tape player sounded like a new one.

The results were the same when the owner installed the tape player back into the car. Then both player and car battery were checked. The player had been connected in opposite polarity in the car. Naturally, with the tape motor running backwards, it would pull tape from the cartridge.

Make & Model: *Automatic Radio Titian 9945*

Symptom: *Excessive crosstalk.*

The tape head would not lift properly. The azimuth and height screws were way out, but this was not the main trouble. A pin at the rear of the head assembly was out and missing. (The pivot pin was found lying under the tape head assembly.)

The tape head was adjusted for height and azimuth settings after the pin was installed.

Make & Model: *Automatic Radio Model CF6-6745*

Symptom: *Intermittent playing.*

Sometimes this tape player would play and other times a tap on one side of the metal case was needed to start tape movement. At first a break in the power lead or motor wiring was suspected. But everything except the motor checked out okay. An exact motor number was ordered for this 8-track player.

Make & Model: *Automatic Radio Model MEL-6740*

Symptom: *Won't change channels.*

This tape player would change tracks automatically, but not manually. Either a defective manual switch or wiring was at fault. A broken momentary manual track-shift switch was found and replaced. You will note that both automatic and manual program switches are wired to the same circuit. When one is operating and the other is not, suspect a defective switch.

Make & Model: *Automatic Radio SEA-6801*

Symptom: *Will not shut off.*

Removing the fuse was the only way to turn the player off. A push-in type cartridge switch was dry and sticking. The switch was removed and washed out. Lubricating and cleanup of cartridge switch solved the no-shutoff problem. The tape head was cleaned up at the same time.

In some cases the cartridge switch may be bent out of line and continue to run or not turn the player on. When cartridge switches are located in the top cover, the small metal spring becomes bent and will not turn the player on. Sometimes a dead player may be caused by a poor lead to cartridge switch.

Make & Model: *Automatic Radio Model MN9949*

Symptom: *Extreme distortion in one channel.*

The owner complained that when he transferred the player to another car it refused to play. When connected to a set of speakers, the right channel was normal but the left channel was distorted. Visual inspection revealed the output

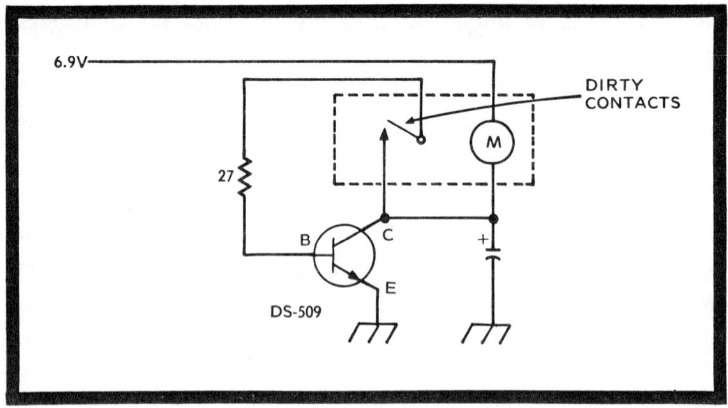

Fig. 12-6. Poor mechanical governor contacts made motor run too fast.

transistor (8P505) metal surface was dented. The power output transistor was removed and was found to be open. Replacing the output transistor with a universal replacement (RCA SK3052) restored the left channel. Undoubtedly, the power transistor was damaged when transferring or connecting power to the other car.

Make & Model: *Automatic Radio Model MNE 6725A*

Symptom: *Strips tapes.*

Sometimes when moving the PC board the right channel would act up. Visual inspection revealed the driver transistor top was blown off. Moving the driver transistor (Q3) with fingers caused the right channel to cut out. Replacing the driver transistor solved the intermittent condition. In this case visual inspection revealed the defective component. Use the scope with test cartridge to signal-trace an intermittent transistor stage in a difficult amplifier section.

Make & Model: *Automatic Radio SCE-6804*

Symptom: *No lights—no motion.*

In this model there were two cartridge switches to turn on motor and amplifier, respectively. The amplifier had a rush noise with volume control wide open in both channels. Since the motor was dead and no channel lights, the other cartridge switch was checked with ohmmeter. The switch continuity was good except the small leaf that pushes against the stereo cartridge was broken off. If a new switch is not readily available, the small leaf can be overlapped and soldered. This is only a temporary repair and should be replaced when a new switch is available.

Make & Model: *Chevrolet model 7305441*

Symptom: *Intermittent operation.*

Amplifier rush noise was indicated in this tape player, but there was no motor rotation. When the motor pulley was started the motor would take off. Sometimes the unit would play all day long. Then the next day the motor would not start. Removing the drive belt would let the motor rotate without a load. The motor bearings were oiled and the results were the same. Replacement of the small motor solved the intermittent problem.

Make & Model: *Chevrolet Model 01BFMT2*

Symptom: *Runs too fast.*

Sometimes when the owner would push the program button to get another music selection, the tape player would respond by changing speeds. The auto stereo tape player was pulled and connected at the test bench. Now the motor appeared to be rotating twice the normal speed. Adding a drag to the motor by fingering the motor pulley returned the motor to normal operation.

A power transistor is found in a simple motor speed circuit of this player (Fig. 12-6). The power transistor tested good. Both resistor (27 ohms) and electrolytic capacitor (250 μF) were normal. The small motor was dismantled and the mechanical governor switch (inside motor) cleaned. The problem was solved.

Make & Model: *Craig Model 3123*

Symptom: *Clicking sound—no music.*

A rush noise indicated the amplifier sections were functioning. The small drive motor would not rotate. A voltage check at motor terminals revealed a defective motor. The motor bearings were cleaned and lubricated. Sometimes the motor would start and other times it would have to be started by hand. The motor had a flat spot and was replaced.

Make & Model: *Craig Model 3108A*

Symptom: *Cartridges jam.*

In this tape player the cartridge seemed not to seat properly. Before removing from auto, the cartridge assembly was inspected and was found jammed with a set of car keys. Removing the keys would let the cartridge seat properly. The tape heads were cleaned up with alcohol and tape cleaning cartridge. You'll find gum wrappers, keys, cigarette butts, and rubber bands in these cartridge containers. And don't forget to look for tape pieces wound around capstan drive assembly.

Make & Model: *Craig Model 3112*

Symptom: *Intermittent speed.*

Sometimes this tape player would run for hours and then speed up. Since no speed circuit is found in this player the motor was replaced with exact part number.

Make & Model: *Craig Model 3116*

Symptom: *Garbled music.*

First, check to see if music is distorted or if there is crosstalk; try a new cartridge. If the results are the same, inspect the tape head. A dirty, oxide-caked tape head can cause weak and distorted output conditions. Excessive crosstalk was noted in this player. A cleanup and tape head alignment solved the garbled music condition.

Make & Model: *Crysler Model 73773*

Symptom: *Wow conditions.*

In this particular model the whole motor assembly will pull out from the radio section. The capstan–flywheel assembly was loose. Before the capstan–flywheel assembly could be tightened, the flywheel was removed. Now the two mounting screws were tightened up. Before the whole assembly was replaced, the tape head and capstan bearings were cleaned and lubricated. Tape head adjustments and firmly mounting the whole capstan–flywheel assembly removed the wow and crosstalk problem.

Make & Model: *Dodge Model 73773*

Symptom: *Will not change channels—tears tapes.*

Sometimes there may be more than one trouble found in players. Excessive oxide dust caused the tearing of tapes. In fact, the tape head and capstan drive shaft were literally packed with tape oxide. Several pieces of tape were wrapped around capstan drive and head ratchet assembly. The excess pieces of tape were lodged against head ratchet and would not let it seat properly. This caused erratic program channel shifting. When a packed oxide capstan drive is found, remove it from the assembly and clean with alcohol. You can clean up and remove excess tape with capstan assembly removed.

Make & Model: *Dynatronics Model S-70*

Symptom: *Grinding sound.*

The flywheel was dragging against the capstan support. Since the shaft is made from very hard steel, it's impossible to pin the flywheel assembly. A layer of epoxy cement around bottom side of flywheel and capstan shaft solved the problem. If the flywheel is not bonded to the capstan drive, the flywheel will drop down again eventually and grind away. Replace the flywheel—capstan drive assembly if the flywheel hole is too large.

Make & Model: *Ford Model T8MS*

Symptom: *Cartridge release won't hold.*

The release solenoid was found burned in this auto tape player. Generally, a jammed solenoid lever or defective eject circuit will cause the solenoid to overheat. After a new solenoid coil was installed, the lever assembly seemed to be normal. But what caused the problem in the first place?

A visual check of the eject transistor circuit revealed a burned resistor (Fig. 12-7). Undoubtedly, a transistor was leaky or shorted. Q1 was removed and tested leaky. Q2, checked with an in-circuit transistor tester, was found to be

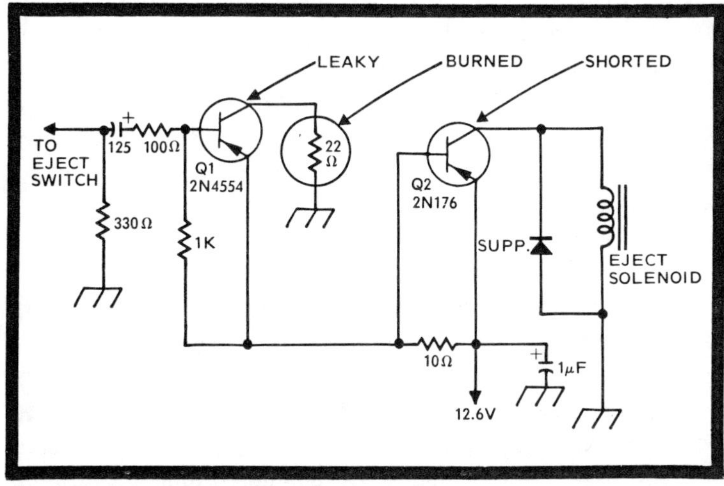

Fig. 12-7. A defective eject circuit, as evidenced by several faults.

shorted. Both transistors and the 22-ohm resistor were replaced. Replacement of defective parts and eject solenoid cured this cartridge release condition.

Make & Model: *Ford Model T75MF*

Symptom: *Intermittent channel.*

The radio section of this tape player was normal, but when a tape cartridge was inserted the left channel appeared intermittent. Sometimes, if the cartridge was jammed or twisted just right, the left channel would function.

A quick check of radio–tape cartridge switch solved the intermittent left channel. The defective switch was cleaned, but the customer returned with the same problem a few weeks later. A new switch was required.

Make & Model: *Ford Model T8MS*

Symptom: *Cartridge will not eject.*

Either the electronic ejection circuit or solenoid were defective in this tape player. The ejection lever assembly was difficult to operate even by hand. Something was binding in the mechanical ejection assembly. The ejection lever that locks the cartridge was bent outward. Alignment of ejection lever assembly solved the cartridge eject system.

Make & Model: *Kraco Model KS-400*

Symptom: *Won't turn off.*

When the tape cartridge was removed, the capstan–flywheel and amplifier would not shut off. In this model the cartridge switch is located at the front and top of cartridge area. The top copper leaf switch was bent down (Fig. 12-8) and would not break contact. A small wheel rotates against cartridge and raises up into position to make contact with top leaf switch in on position alignment. Straightening of cartridge switch solved the turnoff problem.

Make & Model: *Lear-Jet Model A-260*

Symptom: *Improper program indication.*

In this model the program track did not jibe with the selector track indicator. It's best to trip the track-shift switch to program 1. Now rotate track indicator assembly until program 1 indication is centered. Mechanical program indicator can be adjusted in this manner. When electronic dial light indicators are found out of order, suspect a broken or slipping light indicator switch.

Make & Model: *Muntz Model C120*

Symptom: *Too many channel lights.*

Sometimes the channel light indicator assembly would function normally and other times two lights would appear. Also, the numeral 4 was out. The 12V channel light was replaced. The dial light channel switch was found loose on indicator head shaft. These switches can be repaired with epoxy cement if not broken. Replace the entire switch assembly when broken or worn switch contacts are found.

Make & Model: *Muntz Model C170*

Symptom: *No right channel.*

When one channel is found to be defective you can inject an audio signal at the tape head terminal. Signal was found at the left channel terminal, but the right channel was dead. This indicated the tape head was good but the right amplifier section was dead.

The audio signal was injected at the right channel volume control terminal and appeared normal. This indicated the dead channel was between preamp and volume control section. A pencil-type signal tracer located a dead preamp transistor. This *npn* preamp transistor was replaced with a universal transistor replacement.

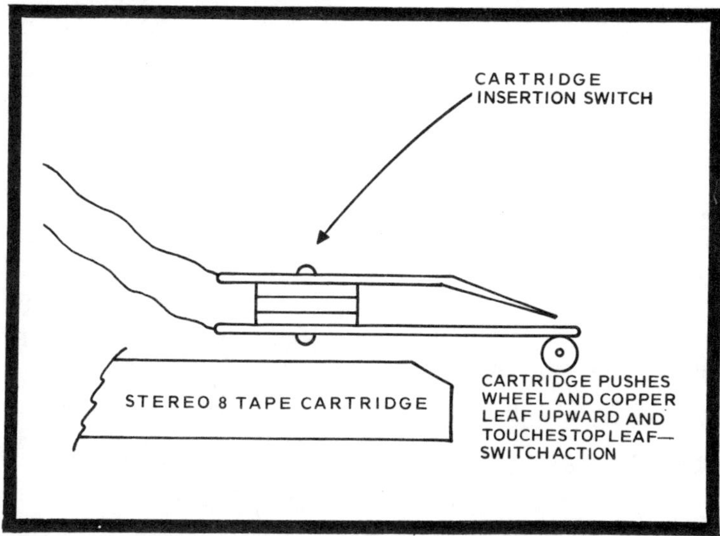

Fig. 12-8. A bent and defective switch leaf prevented the player from turning on.

Make & Model: *Mars Model CS-481*

Symptom: *Intermittent reception.*

In this model the stereo player would act up when tapped or jammed. You could tap the side of metal case and the tape player would quit. The tape player was removed from the car and connected at the repair bench. The results were the same.

Both tape rotation and amplifier were dead at times. Poor connection at one of the power contacts (on PC board) was responsible for the intermittent operation.

Make & Model: *Ranger Stereo 8 player*

Symptom: *Weak right channel.*

When the volume was turned up and balance control rotated to the right channel, this tape player would motorboat. The music was very weak in the right channel. The tape head was checked and cleaned. The results were the same: weak audio.

Sometimes it's difficult to locate the correct pair of output transistors in the amplifier section. You can trace the right speaker wiring back to the amplifier section, but this takes time. Simply place the VTVM lead to the metal case (collector terminal) of output transistors. If the amplifier and output transistor are functioning, you will note a fluctuation of meter movement.

One power output transistor was open and the other showed high leakage.

Make & Model: *Silvertone Model 833.63120*

Symptom: *Intermittent operation and channel light.*

The complaint on this auto stereo tape was that another shop had cleaned the small motor but operation was still intermittent. When mechanical governor switch contacts have been cleaned and the motor speed appears intermittent, replace the motor assembly. You can waste a lot of service time in trying to clean the switch contacts the second time.

HOME-TYPE TAPE DECKS

Make & Model: *Automatic Radio Model HMX-4000*

Symptom: *Intermittent channel change.*

Sometimes in this model the ratchet assembly would not change, and the tape head would remain in one position. The solenoid ratchet would not trip the tape head assembly. The ratchet assembly was very dry and would not seat properly. Cleaning up and lubricating the ratchet assembly restored the channel program condition.

Make & Model: *Automatic Radio Model GES-6394*

Symptom: *Wow.*

A loose belt, oil on capstan drive and dry capstan bearing can produce wow. Sometimes when the cartridge is inserted the tape would slow down and speed up. The wow problem was solved with a new motor drive belt.

Make & Model: *Belair Model 394*

Symptom: *Intermittent program change switch.*

When played in battery operation the solenoid sometimes would not change to another channel. The player worked perfectly with an ac power source. Either the batteries were weak or the solenoid ratchet was not engaging properly. Perhaps the plunger was not fully engaged or pulled in far enough to seat ratchet assembly. Of course, the tape sounded normal in battery operation. Two flashlight cells were found questionable and replaced. The program-change switch worked perfect in battery operation. You should change all batteries when one or two are found weak in a battery operated player.

Make & Model: *Cariole Model 19228*

Symptom: *Lights; no sound.*

In this 8-track stereo combo, the lights were on, but there was no sound. The blown fuse was replaced. After a few seconds, the new fuse was blown, but not before I detected a distorted hum in the left channel.

The right channel output transistor was easily removed and tested; it was good. But the insulator was burned through. Replacement of a new insulator between power transistor and heatsink solved the problem.

Make & Model: *Columbia Model 2692*

Symptom: *Crosstalk and distortion.*

Before tearing into this model the tape head was inspected for excessive packing of tape oxide. The cleanup solved the crosstalk and distorted condition.

Make & Model: *Craig Model 3302*

Symptom: *Tape rotation but no sound.*

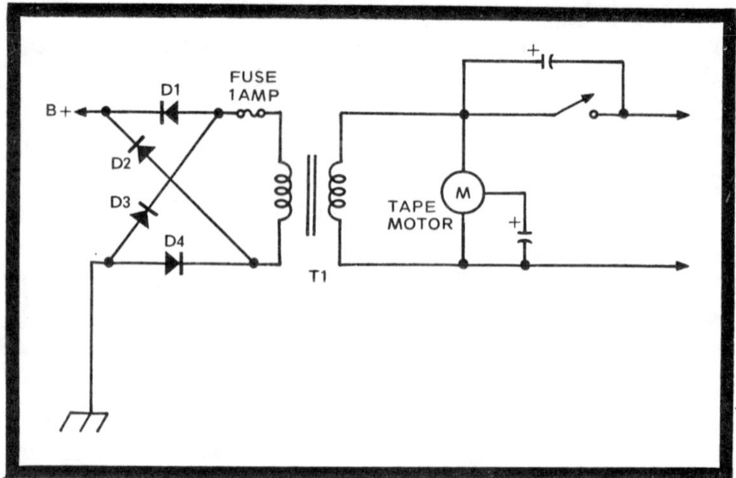

Fig. 12-9. Look for open fuse in secondary winding of power transformer.

Since the player had tape motion and no sound, one might assume the sound problem to be in the amplifier sections. Yet, it was not likely that both amplifier channels were defective. The trouble could be in the power supply circuit, since both sound sections are common to it. Replacement of the fuse (Fig. 12-9) solved the no-sound problem. This fuse is hidden under the printed board at the bottom of player.

Make & Model: *JVC Model CHR-250 UB*

Symptom: *One channel dead.*

This 8-track tape deck had no sound from the left channel. With the volume control full on, I could hear transistor hiss. From this observation the problem should be in the tape head or the preamp transistor stage. Touching a small screwdriver blade to left tape head connection gave out a loud hum. The ohmmeter found the left tape head winding open.

Make & Model: *Lear-Jet Model P529*

Symptom: *Excessive noise—no music.*

When the volume control was turned up I could hear a loud hiss and no music. The motor was not turning and the motor belt was found off of capstan drive. Replacement of belt did not solve the problem. The belt would not stay on since the motor assembly was loose inside shell container.

The motor was removed and examined. The rubber mounting was not tight. Several layers of masking tape solved the problem. After applying masking tape around motor assembly, the unit was installed and tightened into place. A new belt was installed and the motor belt is still turning.

Make & Model: *Lear-Jet Model HA-20*

Symptom: *Defective channel solenoid.*

In this model the program switch would not change to another channel. The solenoid plunger would not engage fully. A layer of wires must be marked and removed to get at the track-shift solenoid.

With the solenoid assembly removed, I could see the plunger would not seat. A plastic core had warped inside the solenoid winding, keeping plunger from engaging fully. The plastic core was drilled out and functioned normally. You should replace the solenoid assembly when you find it burned or excessively worn. Undoubtedly, this solenoid was jammed and had become warm enough to damage the plastic core.

Make & Model: *Lear-Jet Model P529*

Symptom: *Loose belt.*

When operating in the normal position this tape player would throw belts. The belt would only stay on with player on its side. The small flat rubber belt was too loose to stay put and operate. Always clean up capstan—flywheel and motor pulley assembly before replacing motor belt.

Make & Model: *Lloyds Model M614 W*

Symptom: *Continuously blows fuses.*

In this model, the fuse would not hold. A larger fuse was clipped across the defective one and within seconds it, too, went. But in the meantime the right-channel IC began to smoke (Fig. 12-10). Replacement of the IC chip solved the continuous blown fuse problem. The fuse is soldered in and should be replaced with a pigtail type.

Make & Model: *Lloyds Model 2D40*

Symptom: *Intermittent left channel.*

The left channel was intermittent; when operational, it was very weak with distortion. Even when the balance control was rotated to the left channel the volume was very weak. In Fig. 12-11 is a *normal and abnormal voltage* chart of

Fig. 12-10. A shorted IC output stage would keep blowing fuses.

transistors found in the left channel. From the voltage chart, one might suspect Q16 was shorted, since the voltages on all terminals were the same.

Actually, the driver stage (Fig. 12-12), would open, resulting in no emitter voltage. The collector voltage increased almost to supply voltage source. The base voltage

	TRANSISTOR	C	B	E
WHEN DEFECTIVE	Q17	0	+18.5V	+18.75V
	Q16	+18.75V	+18.75V	+18.75V
	Q13	+18.75V	+1V	0V
WHEN NORMAL	Q17	0	+8.2V	+8.4V
	Q16	+19V	+8.6V	+8.4V
	Q13	+8.2V	+.8V	+.2V

Fig. 12-11. Transistor voltage chart.

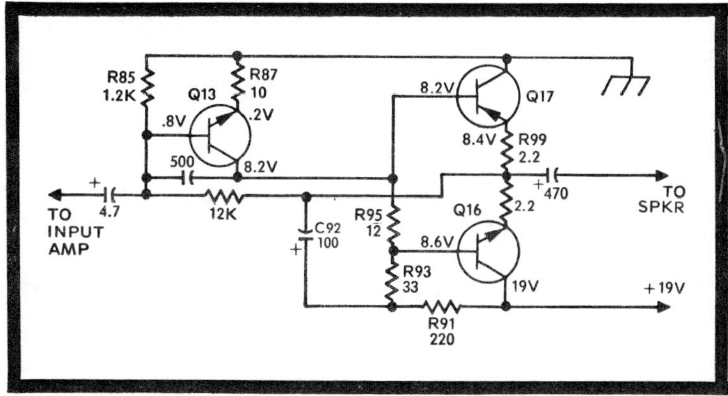

Fig. 12-12. Open Q13 produced the intermittent left channel in this comp-sym circuit.

only reversed to 1 volt. Here the telltale voltage measurements are the collector and emitter terminals. When open, this removed the bias from Q16.

Make & Model: *Lloyds Model 2837*

Symptom: *Intermittent blown fuse.*

In this model the fuse would open up for no apparent reason at all. After several hours of frustration, the problem was traced to the automatic program switch. The wiring of program switch should be changed as shown in Fig. 12-13.

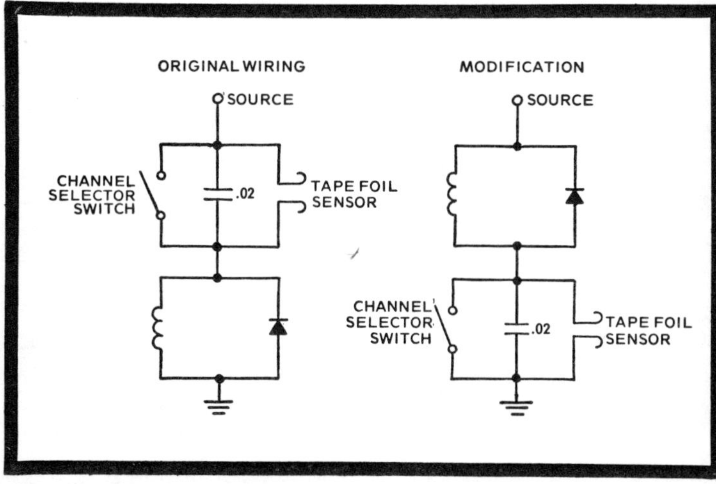

Fig. 12-13. Model 2B37 track change mechanism wiring modification.

The channel switch is wired above the solenoid winding in the original circuit. Simply change the channel selector switch and tape foil sensor to the grounded side of circuit. This track change mechanism wiring modification should be changed in the early models.

Make & Model: *Masterwork Model M508*

Symptom: *Intermittent left channel.*

The radio seemed normal in both channels, though the left channel in tape-play mode was intermittent. The problem was obviously in the tape head or preamp circuits. A pencil noise generator was touched to the left tape head terminal. Very little signal was noted in the speakers. The first preamp transistor was intermittent and replaced.

Make & Model: *Lloyds Model D699*

Symptom: *Erratic balance control.*

At the right side of the balance control the rotation seemed erratic and would not balance with the left channel. A new balance control was installed and the problem was solved. This control is a slide type and must be replaced with exact part number.

Make & Model: *Mayfair Model 484*

Symptom: *Crosstalk.*

Someone had attempted to adjust the tape head in this model and the crosstalk problem still existed. The tape head adjustment is underneath and can be reached with a screwdriver. Correct tape head alignment solved the crosstalk problem.

Crosstalk is most easily noticed on tracks 1 and 4. A test tape should be used in these tape head adjustments. Both height and azimuth tape head adjustments should be made when a new tape head is installed.

Make & Model: *RCA Model YZD 597W*

Symptom: *Slows down in a few minutes.*

This stereo 8 tape deck would slow down within an hour after turn-on. The motor speed could be changed by pushing and pulling on the tape cartridge. This slow-speed problem looked like a dry or gummed up capstan drive bearing.

The capstan—flywheel assembly was removed and its bearing was cleaned and lubed. Also, a new motor belt was installed. This solved the slow-speed problem.

Make & Model: *RCA Model YZB 589E*

Symptom: *Scraping noise.*

On this new stereo 8 tape player a scraping sound could be heard when the capstan—flywheel rotated. Visual inspection revealed the metal-cast flywheel had worked down the capstan drive shaft. Since the tape player was in warranty, a new flywheel assembly was installed.

Make & Model: *Soundesign Model 484*

Symptom: *Intermittent crosstalk.*

Correct tape height adjustment was needed in this stereo tape deck. After tape head adjustment the results were the same. A check against the mechanical drawing indicated that a small tension spring was missing. This small red spring was luckily found on the metal chassis. Replacement of tension tape head spring cured the intermittent crosstalk condition.

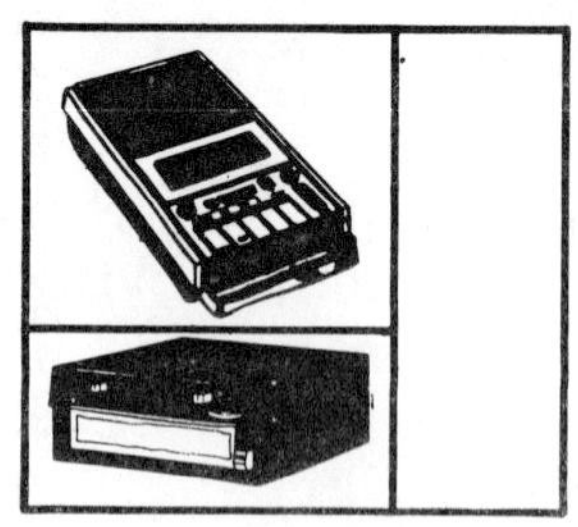

Index

D

E

F

G